Java EE
应用开发实践教程

JAVA EE
YINGYONG KAIFA SHIJIAN JIAOCHENG

涂 祥／著

四川大学出版社

项目策划：梁　胜
责任编辑：曾　鑫
责任校对：孙滨蓉
封面设计：墨创文化
责任印制：王　炜

图书在版编目（CIP）数据

Java EE 应用开发实践教程 / 涂祥著. — 成都 : 四川大学出版社，2019.8
ISBN 978-7-5690-3047-1

Ⅰ. ①J… Ⅱ. ①涂… Ⅲ. ①JAVA 语言－程序设计－高等学校－教材 Ⅳ. ①TP312.8

中国版本图书馆 CIP 数据核字（2019）第 188290 号

书名　Java EE 应用开发实践教程

著　　者　涂　祥
出　　版　四川大学出版社
地　　址　成都市一环路南一段 24 号（610065）
发　　行　四川大学出版社
书　　号　ISBN 978-7-5690-3047-1
印前制作　成都鸿利博雅文化传播有限公司
印　　刷　郫县犀浦印刷厂
成品尺寸　185mm×260mm
印　　张　12.5
字　　数　286 千字
版　　次　2019 年 8 月第 1 版
印　　次　2019 年 8 月第 1 次印刷
定　　价　58.00 元

◆ 读者邮购本书，请与本社发行科联系。
电话：(028)85408408/(028)85401670/
(028)86408023　邮政编码：610065
◆ 本社图书如有印装质量问题，请寄回出版社调换。
◆ 网址：http://press.scu.edu.cn

四川大学出版社
微信公众号

前　　言

本书作为Java EE相关程序设计课程的实践教材，按照计算机方向课程分类分层次教学的整体教改思路设计，并结合编者多年的Java EE及其他程序设计相关课程教学经验编写而成。

全书按照信息管理系统的设计思路为主线，从Java EE基础到热门框架应用，从JDBC数据库编程到持久化框架编程，从单个程序功能设计到整合框架的完整Web数据库应用设计，贴合当前社会实践需求，全面介绍Java EE技术。全书分四个部分共十三个实验，其中在第一部分设计了四个有关Java Web编程基础的实验，主要包括JSP、Java Bean和Servlet编程；在第二部分设计了四个有关Java及Java Web的数据库编程基础的实验，主要包括JDBC、数据库连接池、Hibernate和My Batis实现数据库访问；在第三部分设计了三个有关MVC开发模式的实验，主要包括DAO、Struts、Spring MVC等设计模式和开发框架的实验；在第四部分设计了两个综合应用实验，主要包括Spring容器的IOC依赖诸如和AOP编程应用实验，以及整合Spring、Spring MVC和My Batis框架，实现数据库的Web访问实验，其中还包括Ajax、Site Mesh、Page Helper、Log4j2等技术应用。

本书的实验内容详细、完整，有利于帮助读者学习编程思想、编程技巧和掌握相关技术，并启发读者探索新的编程技术和编程实现方法。

由于编者学识水平有限，书中难免存在疏漏与不足，恳请读者批评指正。

编　者

2018年5月

目　　录

第一部分　Servlet 与 JSP

实验一　JSP 开发基础

一、实验目的

1. 熟悉 Java EE 开发环境。

2. 掌握 JSP 的相关概念、JSP 技术及相关语法。

3. 掌握 JSP 的 9 个隐含对象和 4 种属性范围。

4. 掌握在 JSP 页面共享数据的方法。

二、基础知识

1. JSP 的基本概念。

JSP(Java Server Page)指的是 Java 服务端网页，其页面的构成形式是在传统的 HTML 网页文件中加入了 Java 程序片段和 JSP 标签。

2. JSP 的工作原理。

当用户第一次请求 JSP 页面时，JSP 引擎会通过预处理把 JSP 文件中的静态数据(HTML 文本)和动态数据(java 脚本)全部转换为 Java 代码。比如，对于静态的 HTML 文本只是简单地使用 out. println( )方法包裹起来，而对于 Java 脚本则选择保留或做简单的处理。然后，JSP 引擎将生成的. java 文件编译成 Servlet 类文件(. class)。编译后的 class 对象被加载到容器中，再由 Web 容器(Servlet 引擎)像调用普通 Servlet 程序一样的方式来装载和解释执行这个由 JSP 页面翻译成的 Servlet 程序。

3. JSP 的 9 个隐含对象。

JSP 中存在 9 个隐含对象，是 Web 容器加载的一组类实例，它们不需要显式声明，可直接使用。

表 1－1　JSP 的 9 个隐含对象

隐含对象	所属的类	说明
request	javax. servlet. http. HttpServletRequest	客户端的请求信息
response	javax. servlet. http. HttpServletResponse	传回客户端的响应
session	javax. servlet. http. HttpSession	与请求有关的会话

续表1－1

隐含对象	所属的类	说明
out	javax. servlet. jsp. JSPWriter	向客户端浏览器输出的页面信息
application	javax. servlet. ServletContext	一旦创建就保持到服务器关闭
pageContext	javax. servlet. jsp. PageContext	JSP 页面的上下文，用于访问页面属性
page	java. lang. Object	同 Java 中的 this，即 JSP 页面本身
config	javax. servlet. servletConfig	Servlet 的配置对象
exception	java. lang. Throwable	捕捉一般网页中未捕捉的异常

4. JSP 的 4 种属性范围。

在 JSP 中可通过 page Context、request、session 和 application 这 4 个隐含对象调用 set Attribute( )方法在相应作用域内设置属性。

(1)page 范围。是 page Context 对象设置属性的作用域，在一个页面内保存属性，页面跳转之后无效。

(2)request 范围。是 request 对象设置属性的作用域，在一次服务请求范围内有效。注意，在服务器跳转后依旧有效。

(3)session 范围。是 session 对象设置属性的作用域。在一次会话范围内有效，不管何种跳转都能够使用。

(4)application 范围。是 application 对象设置属性的作用域。在整个服务器上保存，全部用户都能够访问。

在 4 种属性范围内，对属性操作的方法有：

(1)设置属性：public void set Attribute(String name，Object attribute)

(2)取得属性：public Object get Attribute(String name)

(3)删除属性：public Object remove Attribute(String name)

三、实验步骤

1. 下载并解压安装 JDK。在官网下载最新的 JDK 版本并安装。如图 1－1 所示。

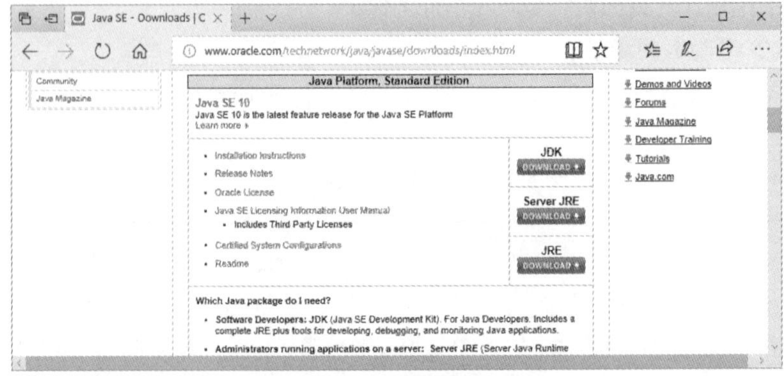

图 1－1　下载安装 JDK

2. 下载并解压安装 Tomcat。在官网下载最新的 Tomcat 版本并安装。如图 1—2 所示。

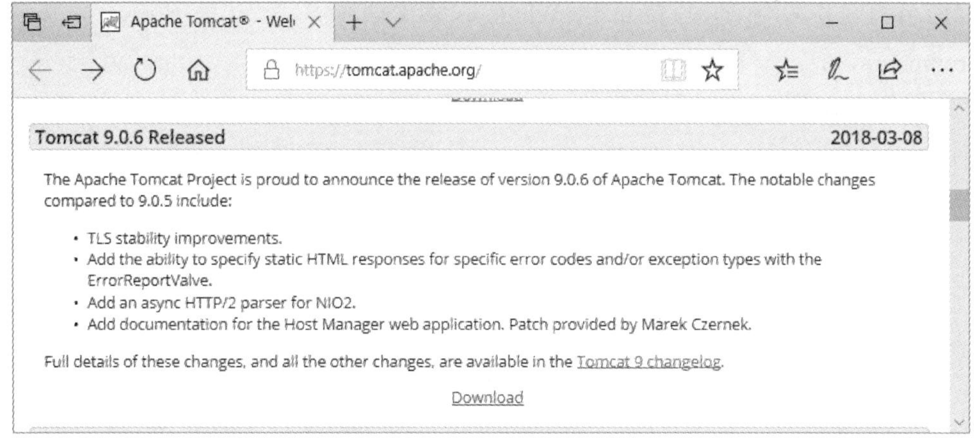

图 1—2　下载并解压安装 Tomcat

3. 下载并解压安装 Eclipse。在官网下载最新的 Eclipse 版本并安装。如图 1—3 所示。

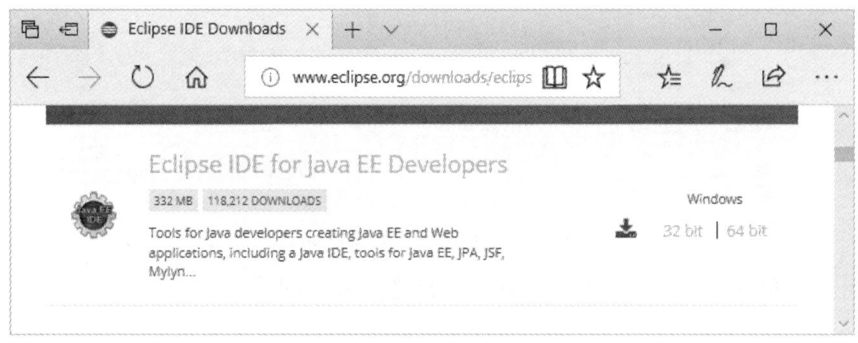

图 1—3　下载并解压安装 Eclipse

需要注意，如果选择 64 位版本的开发环境，则这些软件都应该选择下载 64 位的版本，本教程所使用的版本是 Oxygen. 3 Release(4. 7. 3)。

4. 打开 Eclipse，第一次运行时要求设置工作目录，如图 1—4 所示。

图 1—4　设置工作目录

5. 配置 Java EE 开发环境。

(1) 在 Eclipse 中，点击菜单栏 Windows | Preferences，点击 Server | Runtime Environments，如图 1—5 所示。

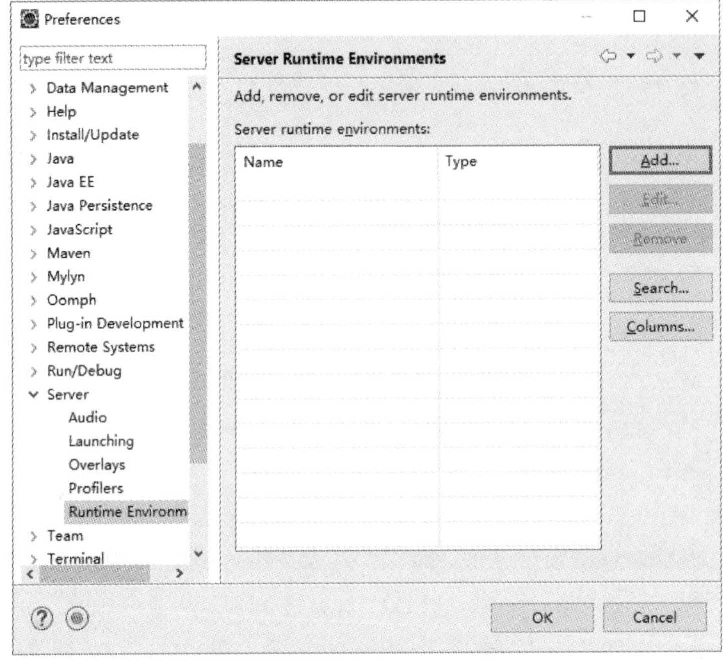

图 1—5　配置服务器运行环境

(2)点击 Add 按钮，新建一个服务器。选择之前安装的 Tomcat 8.0。如图 1—6 所示。

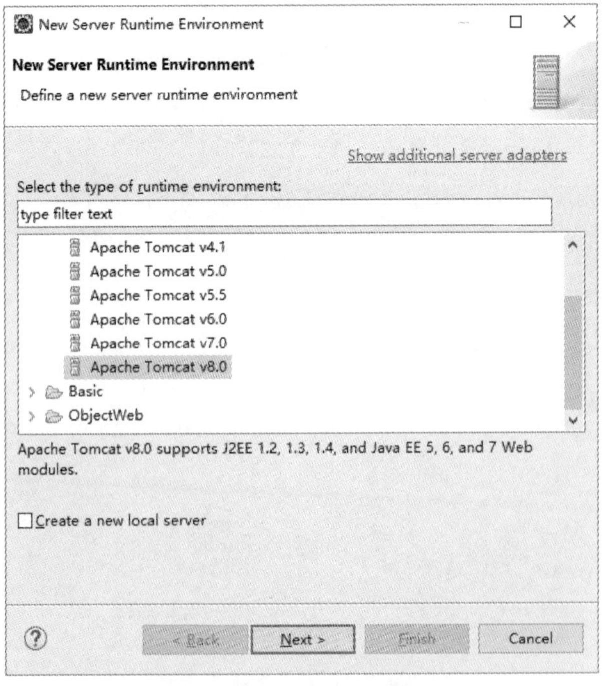

图 1—6　配置 Web 服务器

(3)点击 Next,进入下一步,设置 Tomcat 的安装路径。如图 1-7 所示。最后点击 Finish 完成配置。

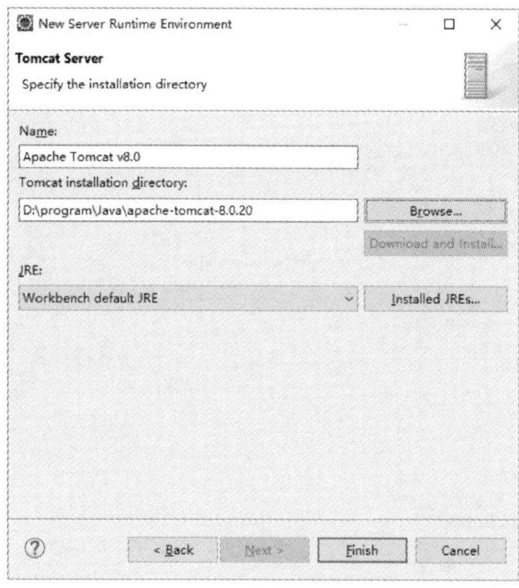

图 1-7 设置 Tomcat 安装目录

6. 设置默认字符集。

在 Eclipse 中,会默认使用当前操作系统的字符集,比如 GBK。在进行 Web 开发时,推荐设置统一字符集 UTF-8。

(1)在 Eclipse 中,点击菜单栏 Windows | Preferences,点击 General | Workspace。如图 1-8 所示。

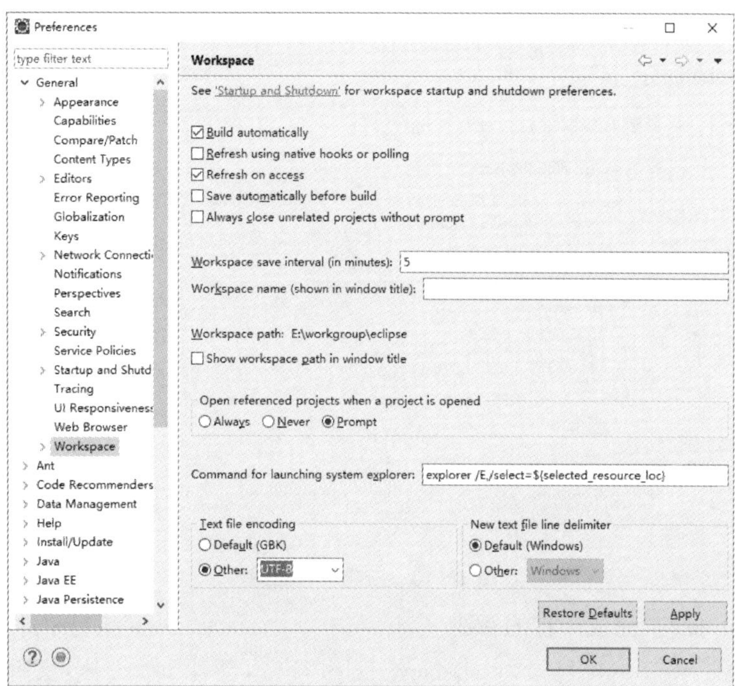

图 1-8 设置字符集

在 Text file encoding 区域选择 Other：UTF－8，设置 Web 开发默认的字符编码为 UTF－8。

(2)在 Eclipse 中，点击菜单栏 Windows | Preferences，点击 Web | JSP File。如图 1－9 所示。

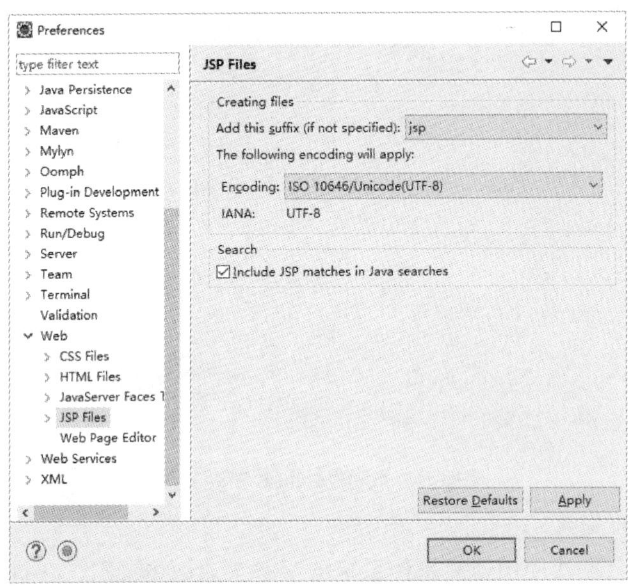

图 1－9　设置 JSP 文件的默认字符集

按照图 1－9 将 Creating files 区域的 Encoding 设置为 UTF－8。

7. 新建一个 Web 工程，通过表单组件传递参数。

(1)在 Eclipse 中，点击菜单栏 File | New | Dynamic Web Project，新建一个 Web 工程 p01_1，如图 1－10 所示。

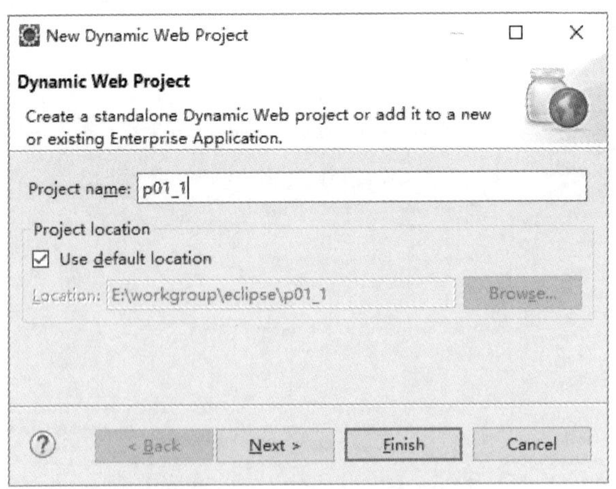

图 1－10　新建 Web 工程

点击 Finish 按钮，完成工程的新建。

(2)右击 Web Content 目录，点击 New | JSP File，新建 login. jsp 文件。如图 1－11 所示。

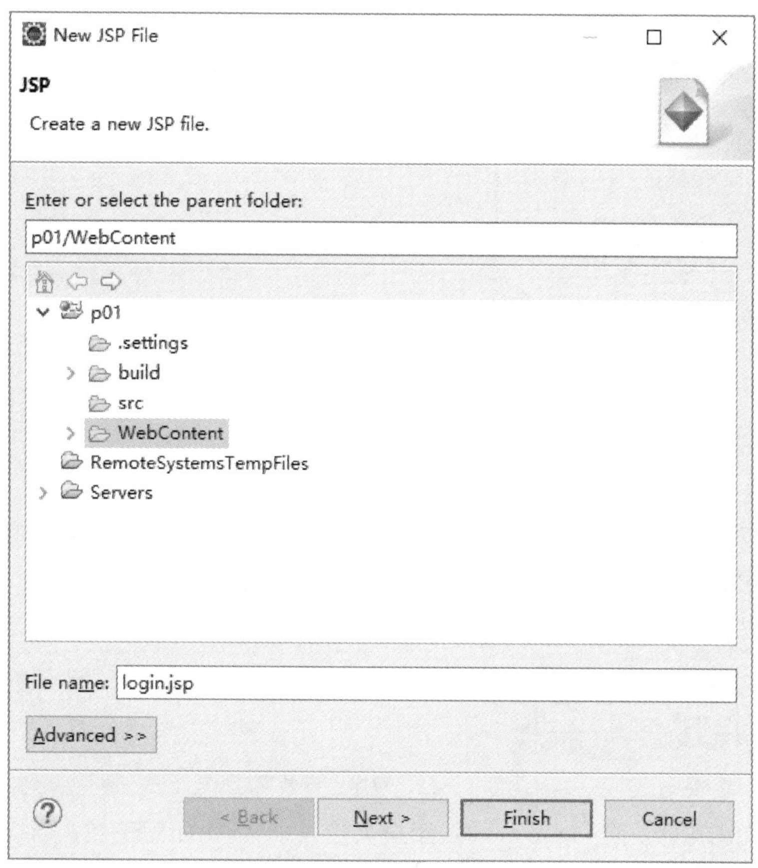

图 1—11　新建 JSP 文件

点击 Finish 按钮，完成 JSP 文件的新建。

(3)双击 login. jsp 文件，进入编辑状态，输入如下所示代码。

```
<%@ page contentType="text/html; charset=UTF-8"pageEncoding="UTF-8"%>
<html>
<head>
<title>登录页面</title>
</head>
<body>
<form action="welcome. jsp" method="post">
用户名:<input type="text" name="yhm"><br>
密码:<inputtype="password" name="mm"><br>
<input type="submit" value="登录">
</form>
</body>
</html>
```

(4)用相同方法新建 welcome. jsp 文件,并输入如下所示代码。

```
<%@ page contentType="text/html; charset=UTF-8" pageEncoding="UTF-8"%>
<html>
<head>
<title>欢迎登录</title>
</head>
<body>
欢迎用户"<%=request. getParameter("yhm")%>"登录系统!
</body>
</html>
```

(5)选中 login. jsp 文件,点击菜单栏 Run | Run As | Run On Server,弹出对话框选择确定,可以在 Eclipse 的内置浏览器上看到运行效果。如图 1—12 所示。

图 1—12　程序运行效果

(6)尝试在浏览器访问 login. jsp 时,用户名输入中文,比如“张三”。当提交表单后,页面显示乱码。此时,需要在 welcome. jsp 页面的代码中,添加如下代码:

```
<% request. set Character Encoding("utf-8"); %>
```

则可以解决参数传递时的中文乱码问题。

8. 用 session 对象实现用户登录验证。

(1)在 Eclipse 中,新建一个 Web 工程 p01_2,工程中新建 3 个 JSP 页面 login. jsp、welcome. jsp 和 logout. jsp。

(2)login. jsp 页面的部分代码如下所示。

```
……
<body>
<form action="login. jsp" method="post">
用户名:<input type="text" name="yhm"><br>
密码:<inputtype="text" name="mm"><br>
<input type="submit" value="登录">
</form>
<%String name = request. getParameter("yhm");
String password = request. getParameter("mm");
if (name ! = null && password ! = null) {
```

```
if ("hbmy". equals(name) && "123456". equals(password)) {
session. setAttribute("flag", "ok");
response. sendRedirect("welcome. jsp");
} else {
%>
<h3>登录失败!!! </h3>
<%}}%>
</body>
……
```

(3)welcome. jsp 页面部分代码如下所示。

```
……
<body>
<%
if("ok". equals(session. getAttribute("flag")))
{
%>
<h1>欢迎访问本系统!!! </h1>
<h2><a href="logout. jsp">注销</a></h2>
<%
}
else{
response. setHeader("refresh","2;URL=login. jsp") ;
%>
<h1>您还未登录,请先登录! </h1>
<%}    %>
</body>
……
```

(4)logout. jsp 页面部分代码如所示。

```
……
<body>
<%
session. invalidate( );
%>
<a href="welcome. jsp">welcome. jsp</a>
</body>
```

……

四、实验思考

1. 按步骤完成实验内容,每个步骤截图保存,形成实验报告。

2. 思考实验中 JSP 页面出现中文乱码的原因,及其解决方案。

3. 尝试在第 8 步实验中,不通过 login. jsp 页面,直接访问 welcome. jsp 页面会有什么结果? 体会 session 对象在用户身份验证中的应用原理。

实验二　JSP+ JavaBean 开发模式

一、实验目的

1. 掌握 Java Bean 的定义和使用。

2. 掌握 EL 表达式和 JSTL 标签库的使用。

3. 掌握 JSP+ Java Bean 的 Web 开发模式。

二、基础知识

1. Java Bean 基础。

Java Bean 是一个遵循特定写法的 Java 类。这个 Java 类必须有一个无参构造方法;属性定义为私有;但私有的属性必须可以通过公有的方法进行修改和访问,该公有方法遵守一定的命名规则,通常命名为 set<Name> 和 get<Name> ,这里的 Name 就是 JavaBean 中封装的属性,并且首字母大写。如下例所示。

```
……
public class Students Bean {
private String first Name = null;
private String last Name = null;
private int age = 0;
public Students Bean( ) { }
public String get First Name( ){
return first Name;}
public String get Last Name( ){
return last Name;}
public int get Age( ){
return age; }
public void set First Name(String first Name){
```

```
this. first Name = first Name; }
public void set Last Name(String last Name){
this. last Name = last Name; }
public void set Age(int age) {
this. age = age; }
}
......
```

2. EL 表达式。

EL 即表达式语言(Expression Language),其基本格式为:${表达式}。一般用于显示数据,功能跟<%=表达式%> 一样。EL 是 JSP2. 0 规范的一部分,只要 Web 容器支持 Servlet2. 4/JSP2. 0,就可以在页面中直接使用 EL 表达式。

为了能够获得 Web 应用程序中的共享数据,EL 表达式中定义了 11 个隐含对象,见表 2—1。

表 2—1 EL 的 11 个隐含对象

隐含对象	对象类型	说明
page Context	javax. servlet. jsp. PageContext	用于访问 JSP 内置对象
param	java. util. Map	包含页面所有参数的名称和值的集合
paramValues	java. util. Map	包含页面所有参数的名称和多个值的集合
header	java. util. Map	包含每个 header 名和值的集合
headerValues	java. util. Map	包含每个 header 名和可能的多个值的集合
cookie	java. util. Map	包含每个 cookie 名和值的集合
initParam	java. util. Map	包含 Servlet 上下文初始请求参数名和值的集合
pageScope	java. util. Map	包含 page 页面范围内的属性值的集合
requestScope	java. util. Map	包含 request 请求范围内属性值的集合
sessionScope	java. util. Map	包含 session 绘画范围内的属性值的集合
applicationScope	java. util. Map	包含 application 应用范围内的属性值的集合

3. JSTL 标签。

JSTL 指的是 JSP 标准标签库(JSP Standard Tag Library),是一个不断完善的开放源代码的 JSP 标签库。JSTL 支持通用的、结构化的任务,比如迭代、条件判断、XML 文档操作、国际化标签、SQL 标签。在 JSP 页面上引入 JSTL 的引用如下代码所示。

```
<%@taglib uri="http://java. sun. com/jsp/jstl/core" prefix="c" %>
```

使用 EL 和 JSTL 的目的是在软件的分层设计中,尽量避免在 JSP 页面中出现 Java 的逻辑代码。

三、实验步骤

1. 新建一个 Web 工程，使用 JSP 标准动作访问 Java Bean。

(1)在 Eclipse 中，点击菜单栏 File | New | Dynamic Web Project，新建一个 Web 工程 p02_1。在工程 p02_1 中，右击 Java Resources | src 目录，新建一个名为 User 的类。如图 2—1 和 2—2 所示。

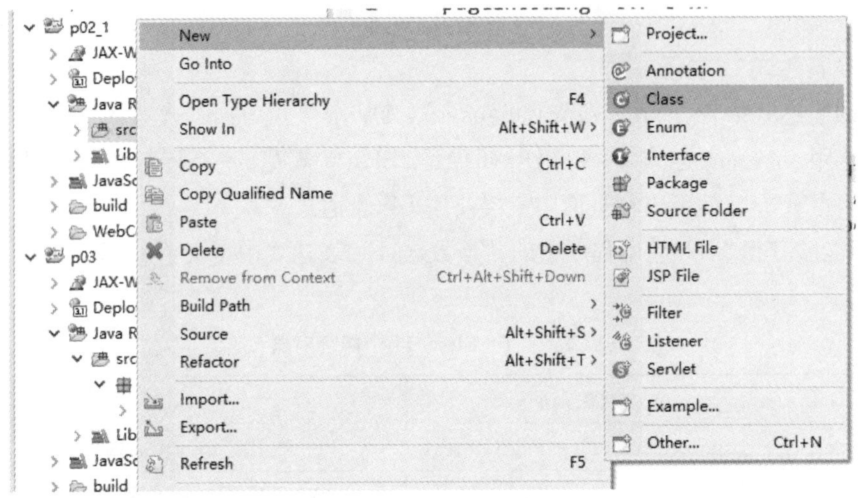

图 2—1　新建一个类

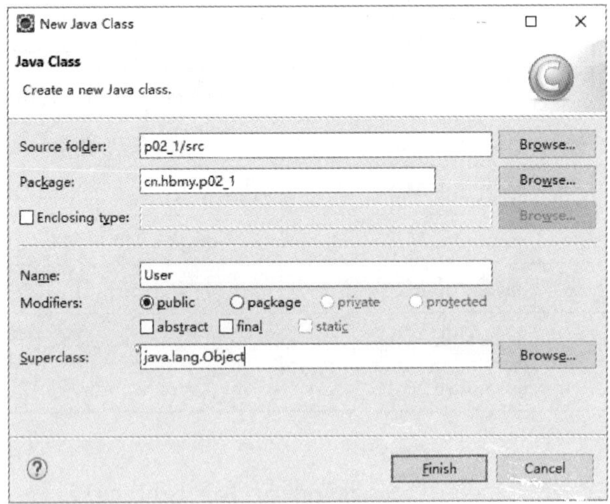

图 2—2　新建 User 类

User 类的代码如下所示。

```
package cn. hbmy. p02_1;
public class User {
private String yhm;
private String mm;
private String yhlx;
```

```
public String getYhm( ) {
return yhm;}
public void setYhm(String yhm) {
this.yhm = yhm;}
public String getMm( ) {
return mm;}
public void setMm(String mm) {
this.mm = mm;}
public String getYhlx( ) {
return yhlx;}
public void setYhlx(String yhlx) {
this.yhlx = yhlx;
}
}
```

说明:可以通过 Eclipse 自动生成 setter 和 getter 方法。具体方法为:先定义类中的属性,然后在代码编辑区域内单击鼠标右键,选择 Source | Generate Setters and Getters…,并做相应设定,将自动生成对应属性的 setter 和 getter 方法。

(2)在工程 p02_1 中添加 2 个 JSP 文件 login.jsp 和 welcome.jsp。其部分代码如下所示。

```
<! --login.jsp-->
……
<body>
<form action="welcome.jsp" method="post">
用户名:<input type="text" name="yhm"><br>
密码:<inputtype="password" name="mm"><br>
用户类型:<select name="yhlx">
<option value="教师">教师</option>
<option value="学生">学生</option>
</select><br>
<input type="submit" value="登录">
</form>
</body>
……
<! --welcome.jsp-->
……
```

```
<body>
<jsp:useBean id="user" class="cn.hbmy.p02_1.User" scope="page"/>
<jsp:setProperty property="*" name="user"/>
<h1>当前登录用户:${user.yhm}</h1>
<h1>输入密码为:${user.mm}</h1>
<h1>用户类型为:${user.yhlx}</h1>
</body>
……
```

2. 新建一个 Web 工程,使用 EL 表达式和 JSTL 完成一个判断是否闰年的 Web 应用程序。

(1)在 Eclipse 中,点击菜单栏 File | New | Dynamic Web Project,新建一个 Web 工程 p02_2。在工程 p02_2 中,右击 Java Resources | src 目录,新建一个名为 IsLeap 的类,类代码如下所示。

```
//IsLeap.java
package cn.hbmy.p02_2;
public class IsLeap {
private int year;
private boolean result;
public int getYear( ) {
return year;}
public void setYear(int year) {
this.year = year;}
public boolean isResult( ) {
return result;}
public void setResult(boolean result) {
this.result = result;}
Public Boolean judge( ){
result=false;
if(this.year%4==0 && this.year%100! =0 || this.year%400==0)
result=true;
return result;
}
}
```

(2)在 p02_2 工程中新建 2 个 JSP 文件 input.jsp 和 output.jsp,JSP 文件的部分代码如图 2—7 和图 2—8 所示。

```
<! --input.jsp-->
……
<body>
<h1>判断闰年</h1>
<form action="output.jsp" method="post">
输入年份值:<input type="text" name="year" />
<input type="submit" value="提交" />
</form>
</body>
……
<! —output.jsp-->
……
<body>
<jsp:useBean id="leap" class="cn.hbmy.p02_2.IsLeap"/>
<jsp:setProperty property="year" name="leap"/>
<%if(leap.judge( )) { %>
<h3><jsp:getProperty property="year" name="leap"/>是闰年! </h3>
<%  } else {
%>
<h3><jsp:getProperty property="year" name="leap"/>不是闰年! </h3>
<% } %>
</body>
……
```

(3)选择 input.jsp 页面,点击菜单栏 Run | Run As | Run On Server,弹出对话框选择确定,可以在 Eclipse 的内置浏览器上看到运行效果。如图 2—3 所示。

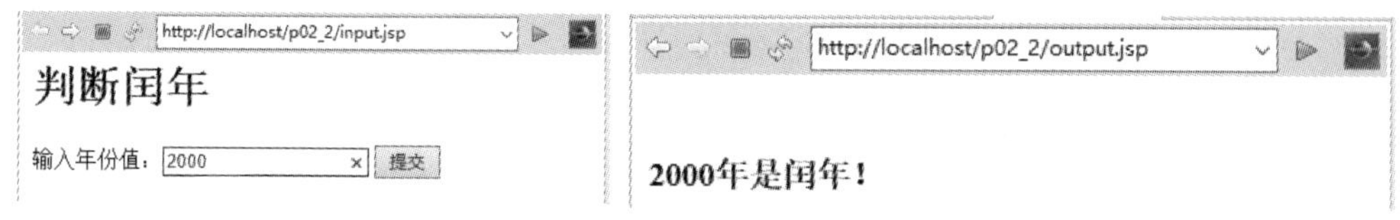

图 2—3 工程 p02_2 运行效果

(4)为了能够使用 JSTL 标签,需要下载标签库 jstl.jar 和 standard.jar,并复制到 WEB—INF 文件夹下的 lib 目录中。然后在 web.xml 里添加 taglib 标注。例如,下面代码添加了 JSTL 的核心标签库。

```
……
<jsp-config>
    <taglib>
```

```
    <taglib-uri>http://java.sun.com/jstl/core</taglib-uri>
     <taglib-location>/WEB-INF/c.tld</taglib-location>


  </taglib>
<taglib>
<taglib-uri>http://java.sun.com/jstl/core-rt</taglib-uri>
<taglib-location>/WEB-INF/c-rt.tld</taglib-location>
</taglib>
</jsp-config>
……
```

注意：在无网络的情况下，需要在 web. xml 中引入标签库；在有网络的情况下，可以不必在 web. xml 文件中作如上设置。

(5)将工程中的代码用 EL 表达式和 JSTL 标签代替，同时修改 IsLeap 类，修改后的 output. jsp 页面和 IsLeap 类的完整代码，如下所示。

```
<! --output.jsp-->
<%@ page contentType="text/html; charset=UTF-8" pageEncoding="UTF-
8"%>
<%@ taglib uri="http://java.sun.com/jsp/jstl/core" prefix="c"%>
<html>
<head>
<title>判断闰年</title>
</head>
<body>
<jsp:use Bean id="leap" class="cn.hbmy.p02_2.IsLeap"/>
<jsp:set Property property="year" name="leap" />
<c:if test="${leap.result}">
<h3><jsp:get Property property="year" name="leap" />年是闰年!
</h3>
</c:if>
<c:if test="${! leap.result}">
<h3><jsp:get Property property="year" name="leap" />年不是闰年!
</h3>
</c:if>
</body>
</html>
```

```
//IsLeap.java
package cn.hbmy.p02_2;
public class Is Leap {
private int year;
private boolean result;
public int get Year( ) {
return year;}
public void setYear(int year) {
this.year = year;}
public boolean is Result( ) {
result=false;
if(this.year%4==0 && this.year%100! =0 || this.year%400==0)
result=true;
return result;}
public void set Result(boolean result) {
this.result = result;}
}
```

说明：在 output.jsp 页面中，为了使用 JSTL 标签库，需要使用如下指令来标注。

```
<%@ taglib uri="http://java.sun.com/jsp/jstl/core" prefix="c"%>
```

四、实验思考

1. 按步骤完成实验内容，每个步骤截图保存，形成实验报告。
2. 当 Web 工程 p02_1 运行时，会出现什么问题？该如何解决。
3. 通过实验体会 JSP+ Java Bean 的编程模式，并简要说明其编程思想。

实验三　Servlet 开发基础

一、实验目的

1. 掌握 Http Servlet 的基本概念、工作原理及其相关编程 API。
2. 掌握 Servlet 的配置文件和注解的 2 种部署方式。
3. 掌握运用 JSP+ Servlet 的开发模式。

二、基础知识

1. Servlet 概念。

Servlet 最初是一种动态 Web 资源开发技术，是服务器端的 Java 应用程序，具有独立于平台和协议的特性，它担当了客户请求与服务器响应的中间层。通常我们也把实现了 Servlet 接口的 Java 程序，称之为 Servlet。与传统的从命令行启动的 Java 应用程序不同，Servlet 由 Web 服务器进行加载。当需要开发一个 Servlet 以实现服务器端响应客户端的动态 Web 资源时，通常需要完成以下 2 个步骤。

(1)编写一个实现了 Servlet 接口的 Java 类。

(2)把开发好的 Java 类部署到 Web 服务器中。

2. Servlet 工作原理。

Servlet 程序是由 Web 服务器调用的，当 Web 服务器收到客户端的 Servlet 访问请求时将做如下处理。

(1)Web 服务器检查是否已经装载并创建了该 Servlet 的实例对象。如果是，则直接执行第(4)步，否则，执行第(2)步。

(2)装载并创建该 Servlet 的一个实例对象。

(3)调用 Servlet 实例对象的 init( )方法。

(4)创建一个用于封装 HTTP 请求消息的 HttpServlet Request 对象和一个代表 HTTP 响应消息的 HttpServlet Response 对象，并调用 Servlet 的 service( )方法将请求和响应对象作为参数传递进去。

(5)Web 应用程序被停止或重新启动之前，Servlet 引擎将卸载 Servlet，并在卸载之前调用 Servlet 的 destroy( )方法。

3. Servlet 编程 API。

HttpServlet 有两个重要的 doPost 方法和 doGet 方法，分别用于响应基于 HTTP 的 post 方法和 get 方法的请求。

(1)doGet( )：当客户端通过 HTML 表单发出一个 HTTP GET 请求或直接请求一个 URL 时，doGet( )方法被调用。

(2)doPost( )：当客户端通过 HTML 表单发出一个 HTTP POST 请求时，doPost( )方法被调用。

4. Servlet 部署。

根据 Servle 版本的不同，可用以下两种方式部署 Servlet。

(1)在 Servlet 2.5 规范之前，Java Web 应用的绝大部分组件都通过 web.xml 文件来配置管理。

(2)Servlet 3.0 规范可通过 Annotation 注解的方式来配置管理 Web 组件，因此 web.xml 文件可以变得更加简洁，这也是 Servlet3.0 的重要简化。

三、实验步骤

1. 新建一个 Web 工程，用于展示 Servlet 的整个生命周期。

(1)在 Eclipse 中,点击菜单栏的 File | New | Dynamic Web Project,新建一个 Web 工程 p03_1。注意,在新建工程同时,为该工程自动生成一个用于 Web 应用程序配置的 web. xml 文件,方法是在新建步骤的 Generate web. xml deployment descriptor 处选中复选框,如图 3—1 所示。默认情况下,实验中所使用的 Eclipse 版本在新建 Web 工程时,不默认生成 web. xml 文件。

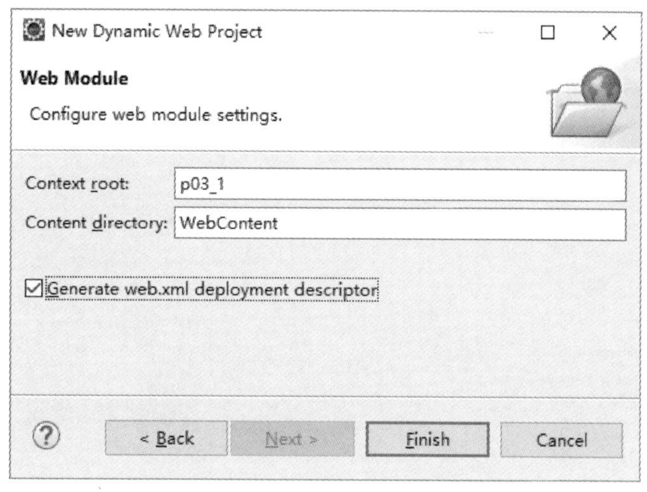

图 3—1　为工程生成 web. xml 文件

(2)为工程 p03_1 添加支持 Servlet 开发的外部库文件 servlet—api. jar。鼠标右击工程 p03_1,在弹出的快捷菜单中点击 Build Path | Configure Build Path 命令。在弹出的窗口中选择 Java Build Path | Libraries 选项卡,单击 Add External JARs 按钮,在弹出的窗口中选择本地 Tomcat 安装目录下的 lib 文件夹中的 servlet—api. jar 库文件,如图 3—2 所示。

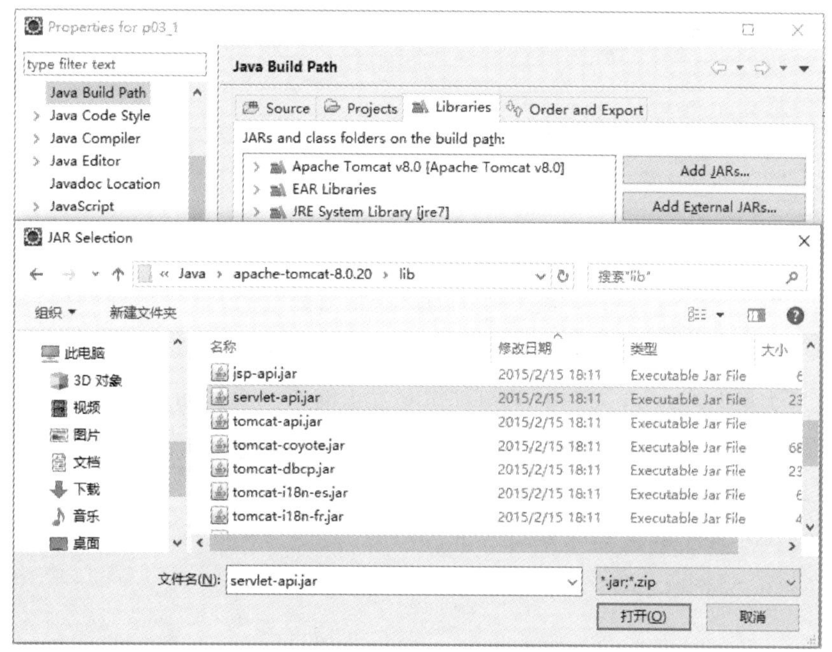

图 3—2　为工程添加 Servlet 库文件

(3)在工程 p03_1 中,右击 Java Resources | src 目录,新建一个名为 LifeCycleServlet 的 Servlet,如图 3—3 和图 3—4 所示。

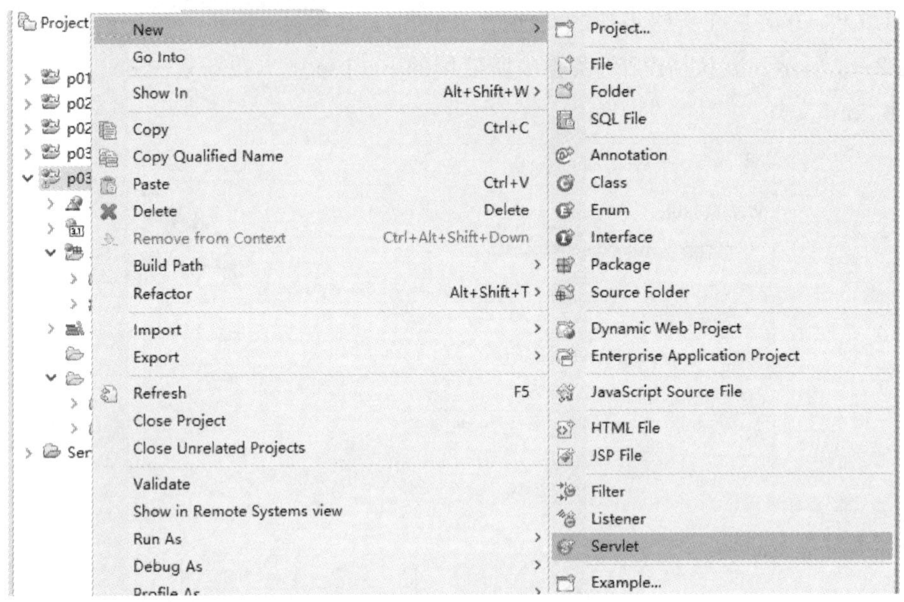

图 3—3　新建 Servlet

Create Servlet

Create Servlet

Specify class file destination.

Project: p03_1

Source folder: /p03_1/src　Browse...

Java package: cn.hbmy.p03_1　Browse...

Class name: LifeCycleServlet

Superclass: javax.servlet.http.HttpServlet　Browse...

☐ Use an existing Servlet class or JSP

Class name: LifeCycleServlet　Browse...

< Back　Next >　Finish　Cancel

图 3—4　设置 Servlet 参数

注意,Servlet 的新建方法和普通类的新建方法略有不同,不同之处在于 Eclipse IDE 帮助代码实现了对 HttpServlet 类的继承和对相应方法的实现。比如,doPost( )、doGet( )等方法的自动实现,以此提高代码编写效率。读者也完全可以通过创建继承 HpptSrvle 类的 Java 类的方式来定义 Servlet,差别仅需要手动添加相关代码而已。

(4)Life Cycle Servlet 类的部分代码如下所示。

```
//Life Cycle Servlet. java
…
public class LifeCycleServlet extends HttpServlet {
```

```
private static final long serialVersionUID = 1L;
public LifeCycleServlet( ) {
super( );
}
public void init(ServletConfig config) throws ServletException {
config. getServletContext( ). setAttribute("data", "* * Servlet 初始化... * *");
}
public void doGet(Http Servlet Request req, Http Servlet Response resp) throws
IOException, Servlet Exception {
req. set Attribute("data", "* * Servlet doGet 处理... * *");
Request Dispatcher rd = req. get Servlet Context( ). get Request Dispatcher("/
output. jsp");
rd. forward(req, resp);
}
public void doPost(Http Servlet Request req, Http Servlet Response resp) throws
IOException, Servlet Exception {
req. set Attribute("data", "* * Servlet doGet 处理... * *");
Request Dispatcher rd = req. get Servlet Context( ). get Request Dispatcher("/
output. jsp");
rd. forward(req, resp);
}
public void destroy( ) {
this. get Servlet Context( ). set Attribute("data", "* * Servlet 销毁...");
}
}
```

(5)在 web. xml 文件中,部署 Servlet,部分代码如下所示。

```
<! --web. xml-->
……
<servlet>
<servlet-name>first Servlet</servlet-name>
<servlet-class> cn. hbmy. p03_1. Life Cycle Servlet </servlet-class>
</servlet>
<servlet-mapping>
<servlet-name>first Servlet</servlet-name>
<url-pattern> /demo</url-pattern>
```

```
</servlet-mapping>
……
```

(7)output. jsp 的部分代码编写如下所示。

```
<! --output. jsp-->
……
<body>
    ${applicationScope. data}
    ${requestScope. data}
    </body>
……
```

这里使用 EL 表达式输出了属性范围内的 data 值。

(8)最后，当在客户端浏览器地址栏输入 http://localhost/p03_1/demo，运行效果如图 3－5 所示。

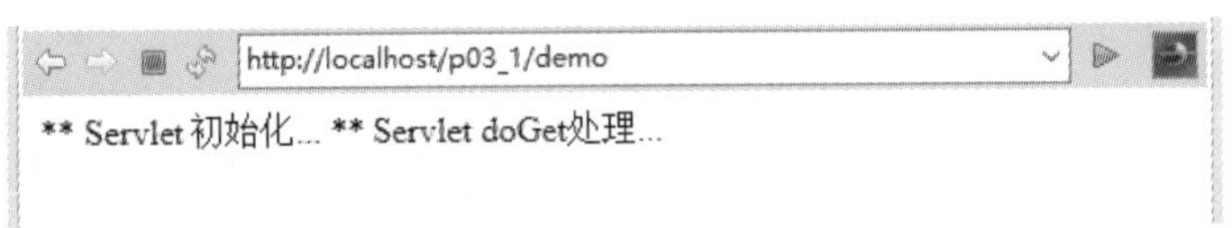

图 3－5　访问 LifeCycleServlet

请读者体会 Servlet 的 url－pattern 和 servlet－class 之间的关系，同时，请尝试用另外一种注解的方式部署 Servlet。注解方式是在 Servlet 类代码中加入如下代码，

```
@WebServlet("/LifeCycleServlet")
```

将该注解代码写在 LifeCycleServlet 类定义的上方。

2. 新建一个 Web 工程，使用 JSP＋ Servlet 方式实现用户登录功能。

(1)在 Eclipse 中，新建一个 Web 工程 p03_2，在工程中新建 3 个 JSP 文件 login. jsp、welcome. jsp 和 fail. jsp，新建 1 个 Servlet 类文件 LoginServlet. java，如图 3－6 和图 3－7 所示。

图 3－6　新建 3 个 JSP 文件

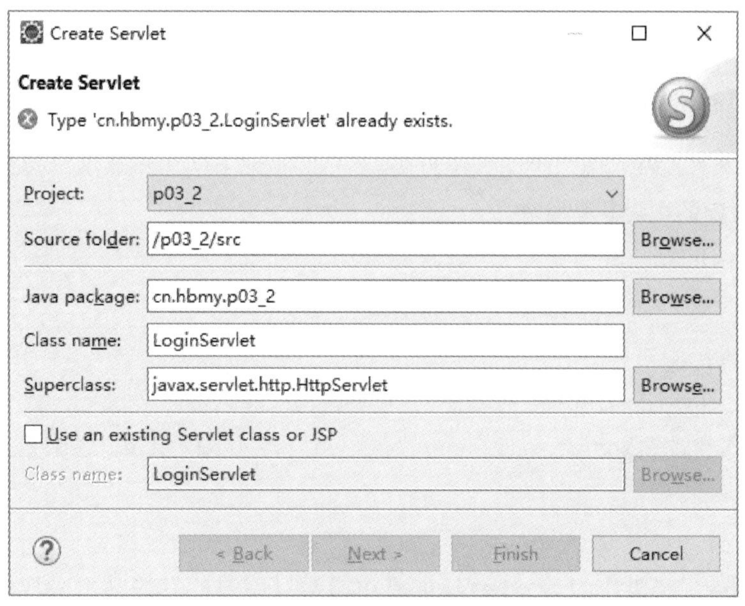

图 3—7 新建 LoginServlet 类

(2)新建的 login. jsp 页面用于用户登录。当用户输入合法用户名、密码和用户类型点击登录按钮后登录到 welcome. jsp 页面,如果登录失败则转入 fail. jsp 页面。Login. jsp 页面的部分代码如下所示。

```
    <! ——login. jsp——>
    ……
<body>
    <form action="login" method="post">
    用户名:<input type="text" name="yhm"><br>
    密码:<inputtype="password" name="mm"><br>
    用户类型:<select name="yhlx">
      <option value="教师">教师</option>
      <option value="学生">学生</option>
    </select>
    <input type="submit" value="登录">
    </form>
    </body>
    </html>
```

welcome. jsp 页面的部分代码如下。

```
    <! ——welcome. jsp——>
    ……
<body>
```

```
欢迎${yhlx}${yhm}登录成功!
</body>
</html>
```

通过 EL 表达式获取用户类型和用户名等信息,实现不同用户显示不同页面内容。

fail.jsp 页面的主要代码如下。

```
<!--fail.jsp-->
……
<body>
登录失败!请点击<a href="login.jsp">这里</a>重新登录!或者5秒后将自动跳转到登录页面!<%response.setHeader("refresh","5;URL=login.jsp");%>
</body>
</html>
```

(3)LoginServlet 类作为 Servlet,其主要功能是对用户的登录请求采取验证操作,并对合法用户根据身份不同,显示不同的欢迎页面内容,而对于非法用户,则自动跳转到 fail.jsp 页面。LoginServlet 类的部分代码如下所示。

```
//LoginServlet.java
……
//请尝试添加注解代码
protected void doGet(HttpServletRequest request, HttpServletResponse response) throws ServletException, IOException {
request.setCharacterEncoding("utf-8");
String yhm=request.getParameter("yhm");
String mm=request.getParameter("mm");
String yhlx=request.getParameter("yhlx");
ServletContext sc=this.getServletContext();
RequestDispatcher rd = sc.getRequestDispatcher("/fail.jsp");
if("hbmy".equals(yhm) && "123456".equals(mm)){
request.getSession().setAttribute("qx", yhlx);
request.setAttribute("yhlx", yhlx);
request.setAttribute("yhm", yhm);
rd = sc.getRequestDispatcher("/welcome.jsp");
}
rd.forward(request, response);
}
……
```

这里使用写入代码的指定用户名和密码来判断用户合法性，读者可思考：在实际应用的 Web 系统中采用将是怎样的方式。

(3)如果在客户端浏览器直接访问 welcome.jsp 页面，会发现能够绕开用户登录，直接访问用户欢迎页面，需要对 welcome.jsp 页面进行如下修改，通过 JSTL 标签进行用户权限判断，从而避免非验证用户访问受限资源。为了使用 JSTL 标签，还需要给 Web 工程添加 JSTL 标签库文件，添加 JSTL 标签库文件的具体操作方法这里不再赘述。

```
//welcome.jsp
<%@ taglib uri="http://java.sun.com/jsp/jstl/core" prefix="c"%>
……
<body>
<c:if test="${empty qx}">
<jsp:forward page="login.jsp"></jsp:forward>
</c:if>
<c:if test="${yhlx=='教师'}">
欢迎教师${yhm}登录成功!
</c:if>
<c:if test="${yhlx=='学生'}">
欢迎学生${yhm}登录成功!
</c:if>
</body>
</html>
```

通过对于 qx 属性值的判断，确定当前访问用户是否为合法用户，若 qx 为空，则为非法用户，页面自动跳转至 login.jsp。根据 yhlx 属性值判断当前用户类型，从而显示不同内容。

(4)选择 login.jsp，点击菜单栏 Run | Run As | Run On Server，在弹出的对话框中选择确定，可以在 Eclipse 的内置浏览器上看到运行效果。如图 3—8、图 3—9 和图 3—10 所示。

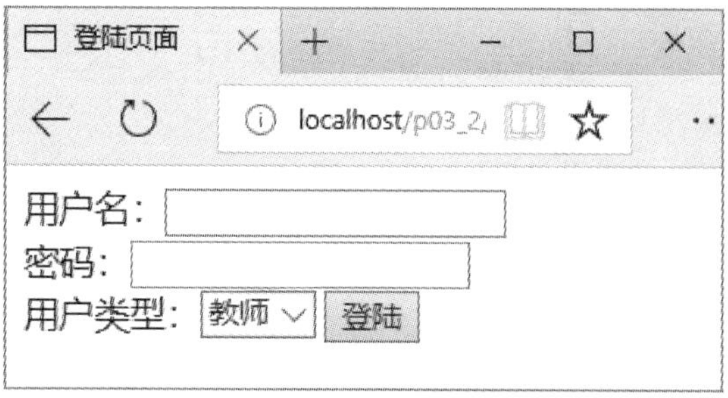

图 3—8　登陆页面

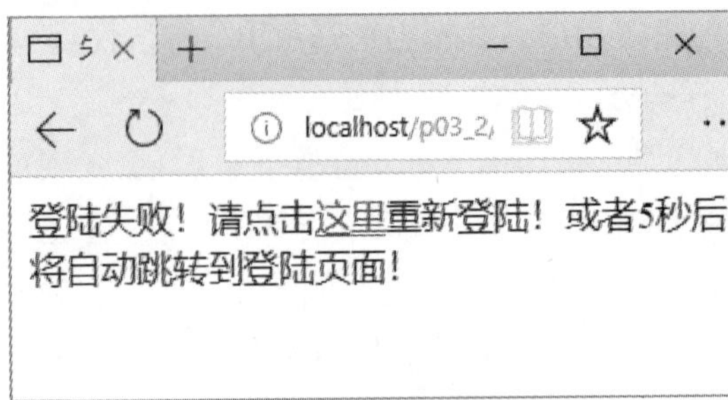

图 3—9　登录失败页面

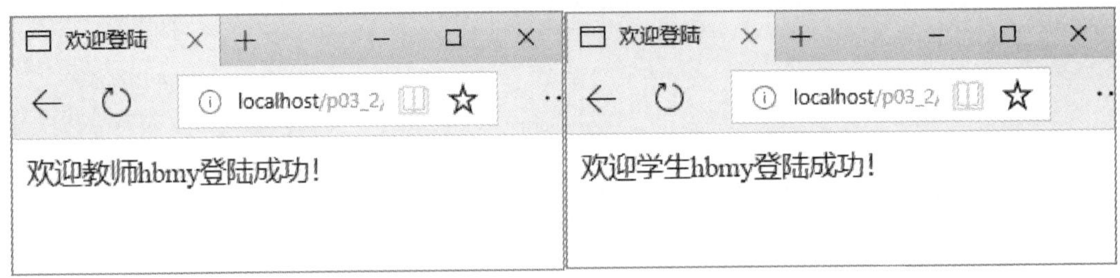

图 3—10　不同用户身份登录页面

四、实验思考

1. 按步骤完成实验内容，每个步骤截图保存，形成实验报告。

2. 请在第 2 个实验中尝试使用注解部署的方式完成 Login Servlet，然后再尝试修改成 web. xml 部署的方式。

3. 请将第 2 个实验中 Servlet 的 forward 跳转方式改为其他方法的跳转，并比较会有什么不同效果？

实验四　Servlet 过滤器和监听器

一、实验目的

1. 理解并掌握 Servlet 过滤器和监听器的概念及工作原理。

2. 掌握 Servlet 过滤器和监听器的编程接口。

3. 掌握 Servlet 过滤器和监听器的两种部署方式。

4. 掌握运用过滤器和监听器开发 Web 应用的方法。

二、基础知识

1. 过滤器概念及工作原理。

Servlet 过滤器(Filter)是在 Java Servlet 规范 2. 3 中定义的小型 Web 组件。主要用于拦截请求和响应，并对请求和响应进行检查和修改。需要注意的是，Servlet 过滤器本身并不产生请求和响应对象，它只能提供过滤作用。Filter 负责过滤的 Web 组件可以是 Servlet、JSP 或者 HTML 文件，即动态或静态的 web 资源。

Filter 总是在资源被调用之前截获请求，检查 ServletReqest，然后根据需要修改 Request 头和 Request 数据；在资源被调用之后截获 ServletResponse，根据需要修改 Response 头和 Response 数据。

具体建立一个过滤器涉及下列 5 个步骤。

(1)建立一个实现了 Filter 接口的类。

(2)在 doFilter 方法中实现过滤行为。

(3)调用 FilterChain 对象的 doFilter 方法。

(4)对相应的 Servlet 或 JSP 页面等注册过滤器。

2. 监听器概念及工作原理。

Servlet 监听器(Listener)也是在 Java Servlet 2.3 中定义的一个 Java Web 组件，是一个实现特定接口的 Java 应用程序。其作用是监听另一个 Java 对象(包括：Servlet Context、Http Session、Servlet Request)的方法调用或属性改变。当被监听对象的方法被调用或属性发现改变时监听器被触发执行。

使用监听器时，通常先创建实现监听器接口的类，然后绑定需要被监听的监听对象，当被监听对象的相应事件发生时，事件监听机制自动触发之前绑定的监听器方法。

3. 过滤器和监听器的部署方式。

同 Servlet 的部署一样，过滤器和监听器的部署也有两种方式，即 web. xml 文件配置方式和 Servlet3. 0 支持的注解的方式。

(1)Filter 的 web. xml 文件配置方式。

```
<filter>
<filter-name>Some Filter</filter-name>
<filter-class>some Package. Some Filter Class</filter-class>
</filter>
<filter-mapping>
    <filter-name>Some Filter</filter-name>
    <servlet-name>Some Servlet</servlet-name>
</filter-mapping>
```

(2)Filter 的注解方式。

```
@Web Filter(urlPatterns={"/*"},
async Supported=true,
dispatcher Types={DispatcherType. REQUEST},
initParams=@WebInitParam(name="param1", value="value1"))
```

(3)Listener 的 web. xml 配置方式。

```
<listener>
<listener－class>listeners. ContextListener</listener－class>
</listener>
```

(4)Listener 的注解方式。

@WebListener

在被监听对象的类定义中完成注解方式部署。

三、实验步骤

1. 新建一个 Web 工程,运用 Filter 实现设置用户请求统一字符编码。

(1)在 Eclipse 中,新建一个带有 web. xml 文件的 Web 工程 p04_1。在工程中新建 2 个 JSP 页面,request. jsp 和 get Param. jsp。在 request. jsp 页面通过 Form 表单提交请求给 get Param. jsp 页面,在 get Param. jsp 页面显示获取的参数值。页面部分代码如下所示。

```
<! －－request. jsp－－>
……
<body>
<form action＝"get Param. jsp" method＝"post">
需要传递的参数:<input type＝"text" name＝"param1"><br>
<input type＝"submit" value＝"提交">
</form>
</body>
……
<! －－getParam. jsp－－>
……
<body>
获取的参数: ${param. param1}
</body>
……
```

(2)运行工程,发现当传递中文参数时会出现乱码,如图 4－1 所示。

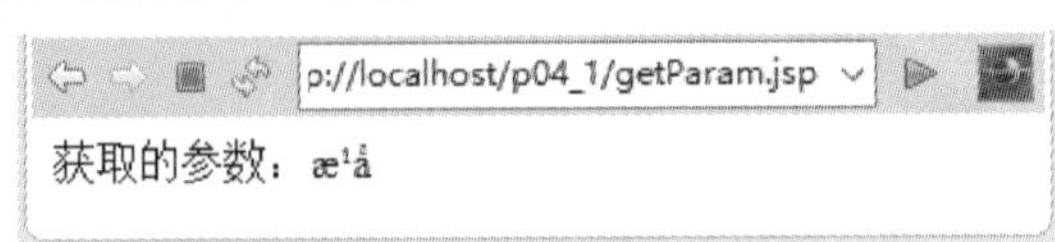

图 4－1 中文参数变成乱码

在实验三中曾使用 request. set Character Encoding("utf－8")方法解决过请求参数的乱码问题,但该方法过于笨拙,需要对所有的 request 请求调用设置字符集方法。本实验中将使用过滤器来解决请求参数的中文乱码问题。

(3)在工程中新建 Filter 类。右击 src 目录，在弹出的快捷菜单中选择 New | Filter，新建 Filter 类，如图 4－2 所示。

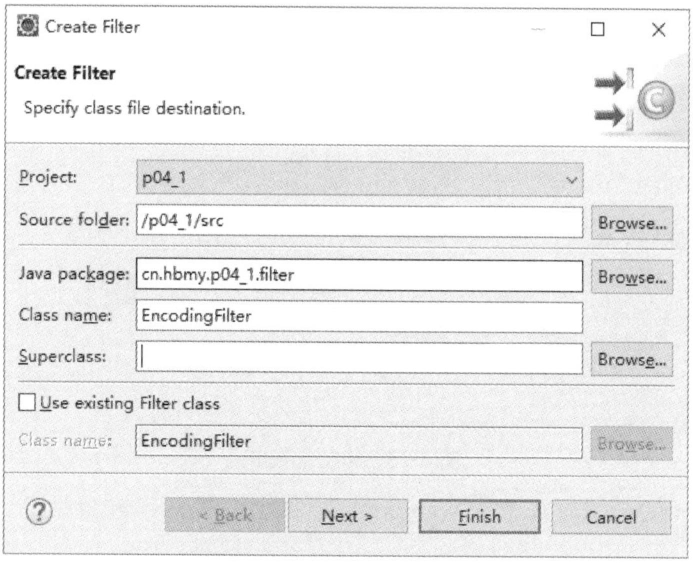

图 4－2　新建 Encoding Filter 类

点击 Finish 按钮，创建 Encoding Filter 类。同 Servlet 类的创建类似，Eclipse IDE 只是帮助实现了 Filter 接口，并继承其相应方法，提高了代码编写效率。所以，也可以通过创建 Java 类的方式新建 Filter。Encoding Filter 类的部分代码如下所示。

```
//Encoding Filter.java
……
@WebFilter(urlPatterns={"/*"})
public class EncodingFilter implements Filter {
……
public void doFilter(ServletRequest request, ServletResponse response, FilterChain chain) throws IOException, ServletException {
HttpServletRequest req=(HttpServletRequest)request;
HttpServletResponse resp=(HttpServletResponse)response;
req.setCharacterEncoding("utf-8");
resp.setCharacterEncoding("utf-8");
chain.doFilter(req, resp);
}
……
```

需要说明的是，该过滤器使用注解方式部署。另外，这种方式的过滤器，以及在 Spring MVC 中添加的 Encoding Filter 过滤器，都只能解决通过 post 方法发送请求的中文乱码，要解决 get 方法的中文乱码需要其他方法，例如，可以在 Tomcat 安装目录下的

server. xml 文件中，相应位置添加 URI Encoding="utf-8"，如下代码所示。

```
<Connector port="80"
protocol="HTTP/1.1"
connection Timeout="20000"
redirectPort="8443"
URIEncoding="utf-8" />
```

(4)运行工程，会发现传递中文参数时的乱码问题已解决，如图 4-3 所示。

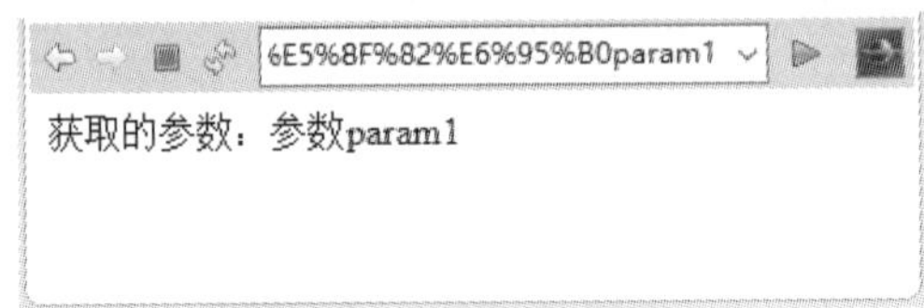

图 4-3　中文乱码解决

(5)将 Filter 的部署方式改成 web. xml 文件配置的方式，配置后的 web. xml 代码如下所示。

```
<! --web.xml-->
......
<filter>
<filter-name>encodingFilter</filter-name>
<filter-class>cn.hbmy.po4_1.filter.EncodingFilter</filter-class>
</filter>
<filter-mapping>
   <filter-name> encodingFilter</filter-name>
<url-pattern>/*</url-pattern>
</filter-mapping>
......
```

这里使用<url-pattern>/*</url-pattern>来设置需要过滤所有资源，还可以通过变换形式，过滤指定包下的资源，或过滤指定的 Servlet 等。

2. 新建一个 Web 工程，运用 Listener 实现统计在线人数的功能。对于在线人数的确认，可以转换成对于 Http Session 对象被创建的确认，当有 Http Session 对象被创建则在线人数加 1，相反，如果有 Http Session 对象被销毁，则在线人数减 1。

(1)在 Eclipse 中，新建一个 Web 工程 p04_2，并新建 session 监听器处理类。右击 src 目录，在弹出快捷菜单中选择 New | Listener，新建一个 Listener，Eclipse 中提供了 Listener 的新建，如图 4-4 所示。

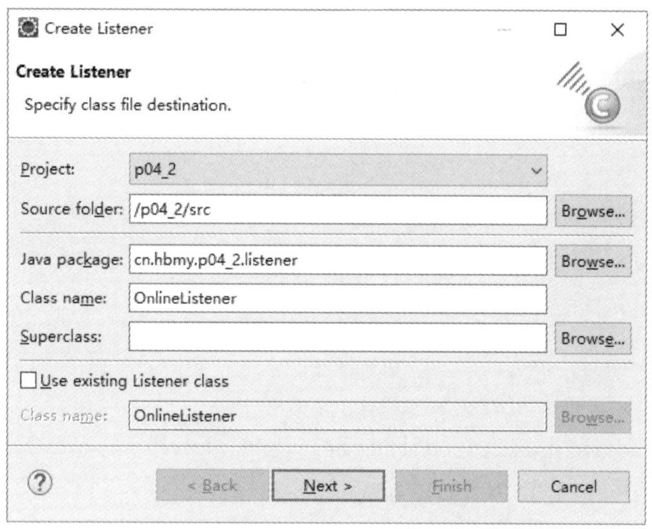

图 4—4 新建 Online Listener 监听器类

点击 Next，弹出对话框用于选择要监听的对象和相应事件，这里选择如图 4—5 所示的 HttpSession 对象的 Changes to Attributes 事件。

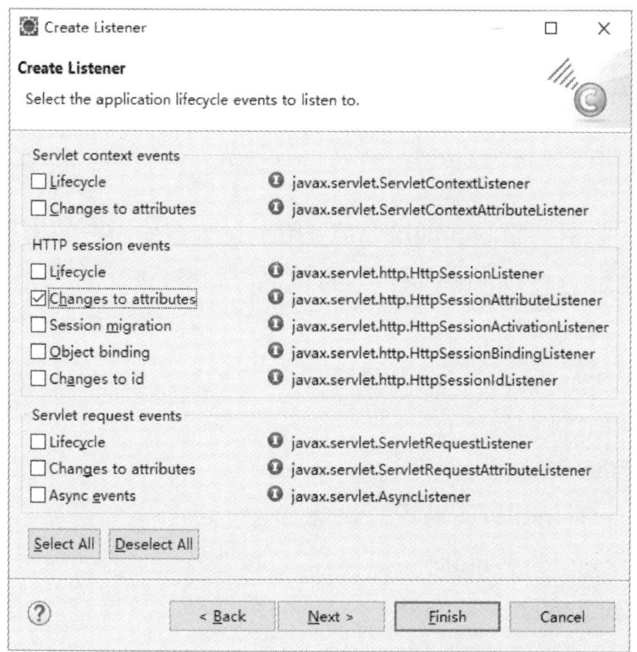

图 4—5 选择监听的对象和事件

点击 Finish，完成新建 Online Listener 监听器实现类。部分代码如下所示。

```
//Onlie Listener. java
……
//注解方式部署 Listener
@Web Listener
```

```
public class Online Listener implements Http Session Attribute Listener {
    public final static String LISTENER_NAME = "_login";
    private static List sessions = new Array List( );
……
    public void attributeRemoved(HttpSessionBindingEvent arg0)  {
        if(LISTENER_NAME. equals(arg0. getName( ))) {
sessions. remove(arg0. getValue( ));
        }
    }
    public void attributeAdded(HttpSessionBindingEvent arg0)  {
        if(LISTENER_NAME. equals(arg0. getName( ))) {
            sessions. add(arg0. getValue( ));
        }
    }
……
    public static List getSessions( ) {
        return sessions;
    }
}
```

(2)在工程 p04_2 中新建 OnlineUser 类,用于封装在线用户的信息,比如用户名、登录时间、用户 IP 地址等。OnlineUser 类的包名 cn. hbmy. p04_2. vo,完整代码如下所示。

```
//OnlineUser. java
package cn. hbmy. p04_2. vo;
public class OnlineUser {
private String ip = null;
private String loginId = null;
private String onlineTime = null;
public OnlineUser(String ip, String loginId, String onlineTime) {
this. ip = ip;
this. loginId = loginId;
this. onlineTime = onlineTime;
}
//省略 setter 和 getter 方法
……
```

```
}
```

(3)在工程 p04_2 中新建 LoginServlet 类,用于模拟登录验证的业务逻辑,当登录验证后的合法用户才能算作系统的在线用户。LoginServlet 类的包名是 cn. hbmy. p04_2. servlet,部分代码如下所示。

```
//Login Servlet. java
……
package cn. hbmy. p04_2. servlet;
……
@WebServlet("/Login Servlet")
public class LoginServlet extends HttpServlet {
……
protected void doGet(HttpServletRequest request, HttpServletResponse response)
throws ServletException, IOException {
String userName=request. getParameter("yhm");
request. getSession( ). setAttribute("_login",
new OnlineUser(request. getRemoteAddr( ),
userName,new Date( ). toString( )));
List sessions = OnlineListener. getSessions( );
request. getSession( ). setAttribute("dataList", sessions);
response. sendRedirect("welcome. jsp");
}
……
}
```

这里只做了象征性的登录验证,只要是通过 login. jsp 访问系统的用户都当做合法用户。

(4)在工程中加载 JSTL 所需要的 jar 包,另外新建 3 个 JSP 页面,login. jsp、welcome. jsp 和 logout. js。login. jsp 用于用户登录;welcome. jsp 用于显示在线用户数;logout. jsp 用于注销当前用户。login. jsp 页面的部分代码如下所示。

```
<! --login. jsp-->
……
<body>
<form action="loginServlet" method="post">
用户名:<input type="text" name="yhm"><br>
<input type="submit" value="登录">
</form>
```

```
</body>
......
```

welcome. jsp 页面代码如下所示。

```
<!--welcome.jsp-->
<%@ page contentType="text/html; charset=UTF-8" pageEncoding="UTF-8"%>
<%@ taglib uri="http://java.sun.com/jsp/jstl/core" prefix="c"%>
<html>
<head>
<title>在线人数</title>
</head>
<body>
<h5>在线人数统计</h5>
<table>
<tr>
<td>No.</td>
<td>ip</td>
<td>登录名</td>
<td>登录时间</td>
</tr>
<c:forEach var="item" items="${dataList}" varStatus="status">
<tr>
<td>${ status.index + 1}</td>
<td>${item.ip}</td>
<td>${item.loginId}</td>
<td>${item.onlineTime}</td>
</tr>
</c:forEach>
</table>
<c:if test="${sessionScope._login! =null}">
<a href="logout.jsp">注销</a>
</c:if>
</body>
</html>
```

logout. jsp 页面代码如下所示。

```
<! --logout.jsp-->
……
<body>
<%session.invalidate( );  %>
已注销,<a href="welcome.jsp">点击这里返回到</a>welcome.jsp!
</body>
……
```

最后运行效果如图 4—6 所示。

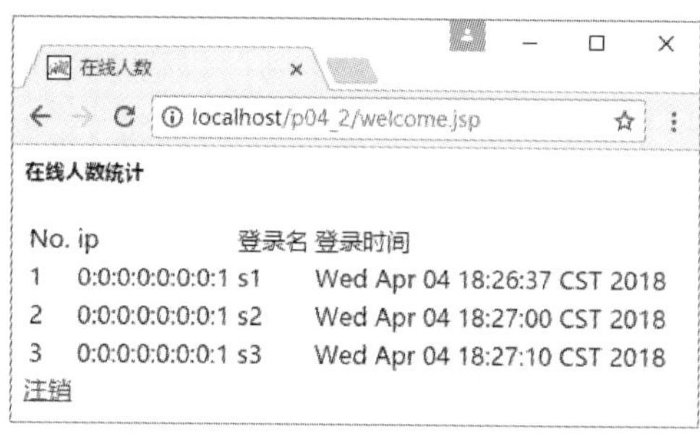

图 4—6 在线人数统计运行效果

四、实验思考

1. 按步骤完成实验内容,每个步骤截图保存,形成实验报告。

2. 比较 Filter、Listener 和 Servlet,简述它们的异同。

3. 在某个 Web 系统中,有 2 个业务需求:(1)需要对用户请求验证其访问权限;(2)需要防止用户重复登录。请根据 Filter 和 Listener 的特点选择并按需求编程,写出编程思路。

第二部分　Java EE 数据库开发

实验五　JDBC 与 Web 数据库编程

一、实验目的

1. 掌握数据库操作的基本 SQL 语句。

2. 理解 JDBC 概念及必要性。

3. 掌握 JDBC 编写 Web 应用程序的基本结构和编程接口。

4. 掌握 CSS 层叠样式表的设计和引用的基本方法。

二、基础知识

1. SQL 与数据库操作。

SQL 即结构化查询语言(Structured Query Language),是用于访问和处理数据库的标准的计算机语言。主要用于数据定义、数据操纵和数据查询,包括数据表结构的定义、查询数据、插入数据、更新数据、删除数据等具体操作。

2. JDBC 概念。

JDBC 即 Java 数据库连接(Java Data Base Connectivity),是一种用于执行 SQL 语句的 Java API,可以为多种关系数据库提供统一访问,它由一组用 Java 语言编写的类和接口组成,使数据库开发人员能够编写数据库应用程序。JDBC 提供了独立于数据库的统一 API,用以执行 SQL 命令。常用的 API 类和接口如下所示。

(1)DriverManager,用于管理 JDBC 驱动的服务类,主要通过它获取 Connection 数据库链接,常用方法如下所示。

public static synchronized Connection getConnection(String url, String user, String password) throws Exception

该方法用于获得 url 对应的数据库连接。

(2)Connection,用于用户与特定数据库建立的连接,只有在连接后才能执行 SQL 语句操作数据库,常用方法如下所示。

①Statement createStatement( ) throws SQLException;

该方法返回一个用于执行具体 SQL 语句 Statement 对象。

②PreparedStatement prepareStatement(String sql) throws SQLException;

该方法返回预编译的 Statement 对象,即将 SQL 语句提交到数据库进行预编译。

③CallableStatement prepareCall(String sql) throws SQLException;

该方法返回 CallableStatement 对象,该对象用于存储过程的调用。

(3)Statement,用于执行 SQL 语句的 API 接口,该对象可以执行数据定义、数据操纵和数据查询等语句,当执行查询语句时返回结果集对象,并可对结果集作进一步操作。常用方法如下所示。

①Result Set execute Query(String sql) throws SQL Exception;

该方法只能用于查询语句,并返回查询结果对应的 Result Set 对象。

②int execute Update(String sql) throws SQL Exception;

该方法用于执行数据操纵语句,并返回受影响的行数。

③boolean execute(String sql) throws SQL Exception;

该方法可以执行任何 SQL 语句,如果执行后第一个结果是 Result Set 对象,则返回 true;如果执行后第一个结果为受影响的行数或没有任何结果,则返回 false。

(4)Prepared Statement,指的是预编译的 Statement 对象,是 Statement 的子接口,它通常允许执行带参数的数据库预编译 SQL 语句,避免对 SQL 语句的重复编译,具有较好性能。由于需要对参数进行设置,所以其常用方法如下。

void set Xxx(int index, value);

根据方法传入的参数类型的不同,需要使用不同的方法。

(5)Result Set 是结果集对象类型,当执行了查询语句时将生成该类型的对象保存结果集。通常需要对查询结果集中的记录做进一步操作,常见的方法如下所示。

①void close( ) throws SQL Exception;

释放、关闭 Result Set 对象。

②boolean absolute(int row);

将结果集移动到第几行,如果 row 是负数,则移动到倒数第几行。如果移动到的记录指针指向一条有效记录,则该方法返回 true。

③void beforeFisrt( );

将 ResultSet 的记录指针定位到首行之前,这是 ResultSet 结果集记录指针的初始状态,记录指针的起始位置位于第一行之前。

④boolean first( );

将 ResultSet 的记录指针定位到首行。如果移动后的记录指针指向一条有效记录,则该方法返回 true。

⑤boolean previous( );

将 ResultSet 的记录指针定位到上一行,如果移动后的记录指针指向一条有效记录,则该方法返回 true。

⑥boolean next( );

将 ResultSet 的记录指针定位到下一行。如果移动后的记录指针指向一条有效记录,则返回 true。

⑦boolean last( );

将 ResultSet 的记录指针定位到最后一行。如果移动后的记录指针指向一条有效记录,则返回 true。

⑧void afterLast( );

将 ResultSet 的记录指针定位到最后一行之后。

3. JDBC 编程步骤。

(1)加载数据库驱动。JDBC 作为数据库访问的规范接口,其具体的实现需要由各个数据库厂商来完成,因此需要加载不同数据库厂商的驱动。

①方法 1：Class. forName("com. mysql. jdbc. Driver")；通过 java. lang. Class 类的静态方法实现。推荐使用该方法。

②方法 2：DriverManager. registerDriver(new com. mysql. jdbc. Driver( ));成功加载后,将 Driver 类的实例注册到 DriverManager 类中。

常用的数据库驱动如下所示。

①Oracle:Oracle. jdbc. driver. OracleDriver

②Mysql:com. MySQL. jdbc. Driver

③SQL Server：com. microsoft. jdbc. sqlserver. SQLServerDriver

④DB2：com. ibm. db2. jdbc. app. DB2Driver

(2)提供 JDBC 连接的 URL。URL 用于标识数据库的位置,通过 URL 地址告诉 JDBC 程序连接哪个数据库,URL 的写法为:

URL=协议名+子协议名+数据源名

常用的数据库 URL。

①Oracle:jdbc:oracle:thin:@machine_name:port:dbname

②Mysql:jdbc:mysql://machine_name:port/dbname

③SQL Server：jdbc：microsoft：sqlserver://machine_name：port；databaseName=dbname

④DB2:jdbc:db2://machine_name:port/dbname

说明:machine_name 指的是数据库所在计算机的名字;port 指的是端口号,其中 Oracle 的默认端口是 1521,MySql 的默认端口是 3306,SQL Server 的默认端口是 1433,DB2 的默认端口是 5000;dbname 指的是要连接的数据库名。

(3)通过 DriverManager 获取数据库的链接。向 java. sql. DriverManager 请求并获得一个 Connection 对象,该对象就代表一个数据库的连接。具体方法如下所示。

DriverManager. getConnection(String url, Stirng user, String pwd)

当使用 DriverManager 来获取连接时需要传入三个参数，url 是 JDBC 连接的 URL，user 是数据库访问的用户名，pwd 是数据库访问的密码。

(4)创建 Statement。要执行 SQL 语句，必须获得用于向数据库发送 SQL 语句的 java. sql. Statement 实例，Statement 实例分为 3 种类型，如下所示。

①Statement stmt = con. createStatement( );

用于执行静态 SQL 语句。通常使用 Statement 实例实现。

②PreparedStatement pstmt = con. prepareStatement(sql);

用于执行动态 SQL 语句。通常使用 PreparedStatement 实例实现。

③CallableStatement cstmt = con. prepareCall("{CALL demoSp(? , ?)}");

用于执行数据库存储过程。通常通过 CallableStatement 实例实现。

(5)执行 SQL 语句。通过调用 Statement 接口提供的 execute、executeUpdate 和 executeQuery 方法来执行 SQL 语句。

(6)处理执行结果。当执行到第(5)步后通常会有两种情况。

①执行更新、插入、删除等操作，返回的是本次操作影响到的记录数。

②执行查询操作返回的是一个 ResultSet 结果集对象。对于结果集对象可作进一步操作。

(7)Jdbc 程序运行完后，切记要释放程序在运行过程中创建的那些与数据库进行交互的对象：如 ResultSet，Statement 和 Connection 对象。特别 Connection 对象，是非常稀有的资源，用完后必须马上释放，如果 Connection 不能及时、正确的关闭，极易导致系统宕机。Connection 的使用原则是尽量晚创建，尽量早释放。资源释放顺序和声明顺序相反，如下所示。

(1)关闭 ResultSet。

(2)关闭 Statement。

(3)关闭 Connection。

3. CSS。

CSS 即层叠样式表(Cascading Style Sheets)，其主要功能是可以对网页元素进行精确控制和格式化，从而达到修饰网页的效果。

三、实验步骤

1. 下载并安装 Mysql。在官网下载 MySQL 的 Community 版本并安装。如图 5—1 所示。

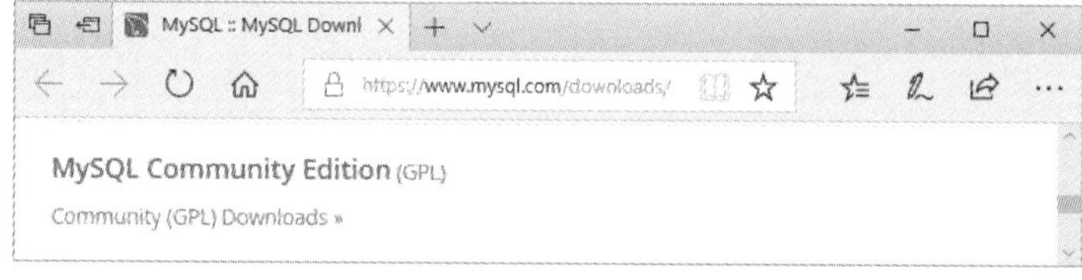

图 5—1 下载安装 MySQL

需要说明的是，MySQL 有许多版本，其中 MySQL Community（社区）版本是开源的，也就是免费的版本，Oracle 对这个版本的 MySQL 不提供技术服务。本课程作为学习 Java Web 数据库编程技术，使用该版本的 MySQL。

2. 通过 Mysql 新建数据库 edu_db，以及新建数据表 edu_stu，其表结构见表 5－1。

表 5－1　数据表 edu_stu 的表结构

字段名称	类型	长度	说明
stuNum	Varchar	9	学号
stuName	Varchar	20	姓名
stuSex	Varchar	1	性别
stuAge	int	2	年龄
stuMajor	Varchar	20	专业

（1）进入命令提示符窗口，输入命令 mysql －u root －p 回车，然后输入密码，回车，进入 mysql 命令行模式。首先建立 edu_db 数据库，输入命令。

```
CREATE DATABASE edu_db;
```

（2）输入如下所示命令，然后回车。建立表 edu_stu 的表结构，并设置 stuNum 字段为主键。

```
mysql>use edu_db;
mysql>CREATE TABLE edu_stu(
->stuNum VARCHAR(9) NOT NULL,
->stuName VARCHAR(20) NOT NULL,
->stuSex VARVHAR(1) NOT NULL,
->stuAge int(2) NOT NULL,
->stuMajor VARCHAR(20) NOT NULL,
->PRIMARY KEY(stuNum)
->) DEFAULT CHARSET=utf8;
```

（3）给 edu_stu 表录入记录。输入类似如下所示命令，然后回车。依次将表 5－2 所示的记录数据插入到 edu_stu 表中。

```
mysql> INSERT INTO edu_stu(
-> stuNum,stuName,stuSex,stuAge,stuMajor)
-> VALUES("031640301","李强","男",19,"信息安全");
```

表 5—2 edu_stu 表记录

学号	姓名	性别	年龄	专业
031640101	张丽	女	20	计算机科学与技术
031640102	李立	男	20	计算机科学与技术
031640103	杨文	男	20	计算机科学与技术
031640201	王琴	女	19	数字媒体
031640202	徐方	男	21	数字媒体
031640203	覃钟	男	20	数字媒体
031640301	李强	男	19	信息安全
031640302	朱雨	女	20	信息安全

3. 新建一个 Web 工程,使用 JSP+ JDBC 的方式实现数据库访问。

(1)在 Eclipse 中,新建一个 Web 工程 p05_1,给该工程添加 My SQL 驱动包,在网上搜索并下载 My SQL 的驱动包,并复制到本工程的 WEB－INF 文件夹下的 lib 目录中。

(2)在工程 p05_1 中新增资源文件 edu_db. properties,其代码如下所示。

```
driver=com.mysql.jdbc.Driver
url=jdbc:mysql://localhost:3306/edu_db
username=root
password=123456
```

说明:本书中的数据库环境设置如上,读者实验时请根据自身具体实验环境设置参数。

(3)在工程 p05_1 中新建一个 JDBC 数据库连接类文件 JDBCUtils. java,如图 5—2 所示。

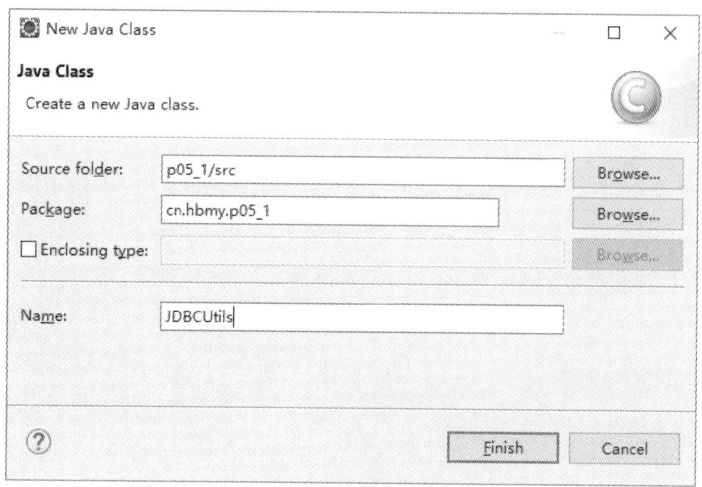

图 5—2 新建 JDBCUtils 类

其部分代码如下所示。

```
//JDBCUtils. java
package cn. hbmy. p05_1;
……
public class JDBCUtils {
private static String driver = null;
private static String url = null;
private static String username = null;
private static String password = null;
static {
try {// 读取 db. properties 文件中的数据库连接信息
InputStream in = JDBCUtils. class. getClassLoader ( ). getResourceAsStream
("edu_db. properties");
Properties prop = new Properties( );
prop. load(in);
driver = prop. getProperty("driver"); // 获取数据库连接驱动
url = prop. getProperty("url"); // 获取数据库连接 URL 地址
username = prop. getProperty("username"); // 获取数据库连接用户名
password = prop. getProperty("password"); // 获取数据库连接密码
Class. forName(driver); // 加载数据库驱动
} catch (Exception e) {
throw new ExceptionInInitializerError(e);
}
}
public static Connection getConnection( ) throws SQLException {
return DriverManager. getConnection(url,username,password);
}
public static void release(Connection conn, Statement st, ResultSet rs) {
if (rs ! = null) {
try {// 关闭存储查询结果的 ResultSet 对象
rs. close( );
} catch (Exception e) {
e. printStackTrace( );
}
rs = null;
```

```
}
if (st ! = null) {
try {// 关闭负责执行 SQL 命令的 Statement 对象
st. close( );
} catch (Exception e) {
e. printStackTrace( );
}
}
if (conn ! = null) {
try {// 关闭 Connection 数据库连接对象
conn. close( );
} catch (Exception e) {
e. printStackTrace( );
}
}
}
}
```

(3)在工程 p05_1 中新建 cn. hbmy. p05_1 包,在包中新建 Student 类,该类的部分代码如下所示。

```
//Student. java
……
package cn. hbmy. p05_1;
public class Student {
private String stuNum;
private String stuName;
private String stuSex;
private int stuAge;
private String stuMajor;
//省略 setter 和 getter 方法
……
```

(4)在工程中 p05_1 新建 Servlet,用于处理显示所有记录。类名为 QueryServlet,包名为 cn. hbmy. p05_1,类的部分代码如下所示。

```
//QueryServlet. java
……
@WebServlet("/QueryServlet")
```

```
……
protected void doGet(HttpServletRequest request, HttpServletResponse response) throws ServletException, IOException {
Connection con=null;
Statement stmt=null;
ResultSet rs=null;
List<Student> dataList=new ArrayList<Student>();
Student stu=null;
try{
con=JDBCUtils.getConnection();
stmt=con.createStatement();
rs=stmt.executeQuery("select * from edu_stu");
while(rs.next()){
stu=new Student();
stu.setStuNum(rs.getString(1));
stu.setStuName(rs.getString(2));
stu.setStuSex(rs.getString(3));
stu.setStuAge(rs.getInt(4));
stu.setStuMajor(rs.getString(5));
dataList.add(stu);}
}catch(Exception e){
e.printStackTrace();;
}finally{
JDBCUtils.release(con, stmt, rs);
}
request.getSession().setAttribute("dataList", dataList);
response.sendRedirect("stu_disAllRec.jsp");
}
……
```

(5)在工程 p05_1 中新建 JSP 文件 stu_disAllRec.jsp,其部分代码如下所示。

```
<!--stu_disAllRec.jsp -->
……
<body>
<form action="QueryServlet" method="post">
<input type="submit" value="查看记录" />
```

```
</form>
<c:forEach var="stu" items="${dataList}">
            ${stu.stuNum}-${stu.stuName}--
            ${stu.stuSex}-${stu.stuAge}-${stu.stuMajor}<br>
</c:forEach>
</body>
</html>
```

(6)选择 stu_disAllRec.jsp 文件,点击菜单栏 Run | Run As | Run On Server,弹出对话框选择确定,可以在 Eclipse 的内置浏览器上看到运行效果。如图 5—3 所示。

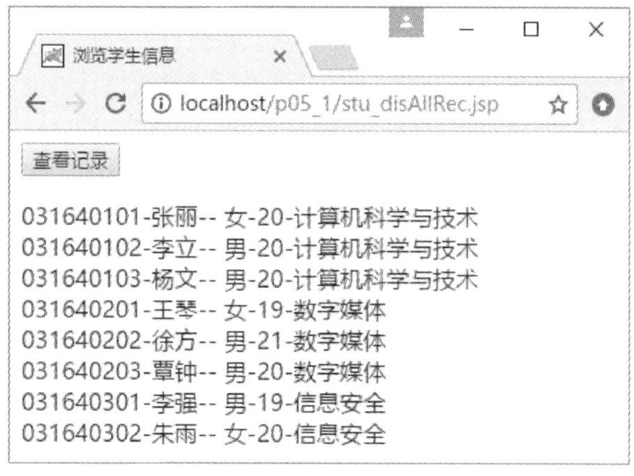

图 5—3　在页面显示 edu_stu 表所有记录

4. 新建一个 Web 工程,使用 CSS 美化上例中页面数据的输出。

(1)在 Eclipse 中,新建一个 Web 工程 p05_2。然后参考上例,为工程加载开发需要的相关 jar 包,并新建和上例相同的 edu_db.properties 属性文件、JDBCUtilse.java 类、Student.java 类、QueryServlet 类和 stu_disAllRec.jsp 文件。

(2)在工程 p05_2 的 WebContent 目录下新建一个 css 目录,用于放置 css 文件。方法如图 5—4 所示。

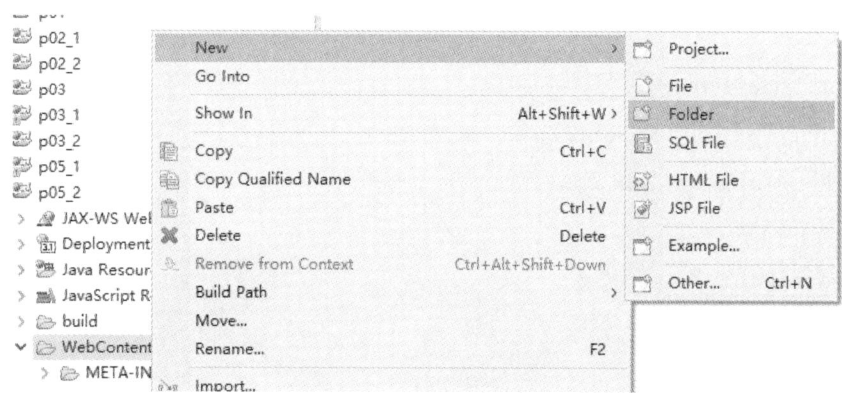

图 5—4　新建 css 目录

点击 Folder 命令后，在弹出对话框中 Folder name 处，输入 css，最后点击确定。

(3)在 css 目录中新增. css 文件。右击 css 目录，弹出快捷菜单中，选择 New | Other，弹出对话框中选择 Web | CSS File，如图 5—5 所示。

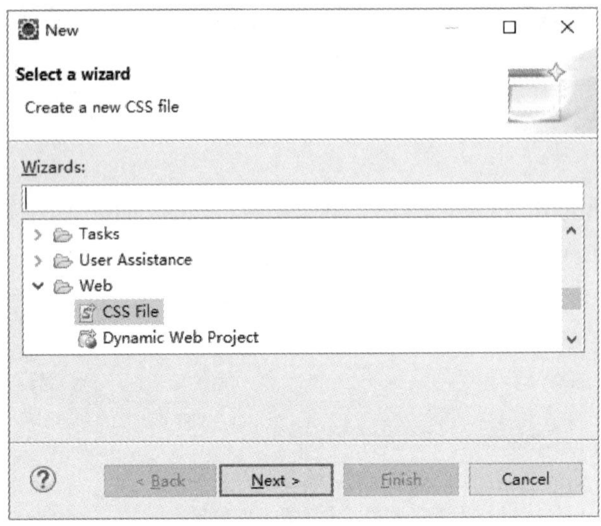

图 5—5　新建 css 文件

点击 Next，给文件命名为 tb. css。

(3)编写 tb. css 代码。其完整代码如下所示。

```
@CHARSET"UTF—8";
#students
  {
  font—family:"Trebuchet MS", Arial, Helvetica, sans—serif;
  width:100%;
  border—collapse:collapse;
  }
#students td, #students th
  {
  font—size:1em;
  border:1px solid #98bf21;
  padding:3px 7px 2px 7px;
  }
#students th
  {
  font—size:1. 1em;
  text—align:left;
```

```
  padding-top:5px;
  padding-bottom:4px;
  background-color:#A7C942;
  color:#ffffff;
  }
#students tr.alt td
  {
  color:#000000;
  background-color:#EAF2D3;
  }
```

(4)选择 stu_disAllRec.jsp 文件，修改文件代码。首先在<head>标签处导入 css 样式表文件，导入方法如下所示代码。

```
<link rel="stylesheet" type="text/css" href="css/tb.css">
```

需要注意的是，此处样式表文件引入的地址是该 css 文件相对于要引用的 jsp 文件的地址，如相对地址"css/tb.css"。请读者思考如果需要改成绝对地址，该如何表示？

然后修改部分代码，如下所示。

```
<!--stu_disAllRec.jsp-->
……
<body>
<form action="QueryServlet" method="post">
<input type="submit" value="查看记录" />
</form>
<table id="students">
<tr>
<th>学号</th>
<th>姓名</th>
<th>性别</th>
<th>年龄</th>
<th>专业</th>
</tr>
<c:forEach var="stu" items="${dataList}">
<tr class='alt'>
<td>${stu.stuNum}</td>
<td>${stu.stuName}</td>
<td>${stu.stuSex}</td>
```

```
<td>${stu.stuAge}</td>
<td>${stu.stuMajor}</td>
<tr>
</c:forEach>
</table>
</body>
……
```

(5)选择 stu_disAllRec.jsp 文件，点击菜单栏 Run | Run As | Run On Server，弹出对话框选择确定，可以在 Eclipse 的内置浏览器上看到运行效果。如图 5—6 所示。

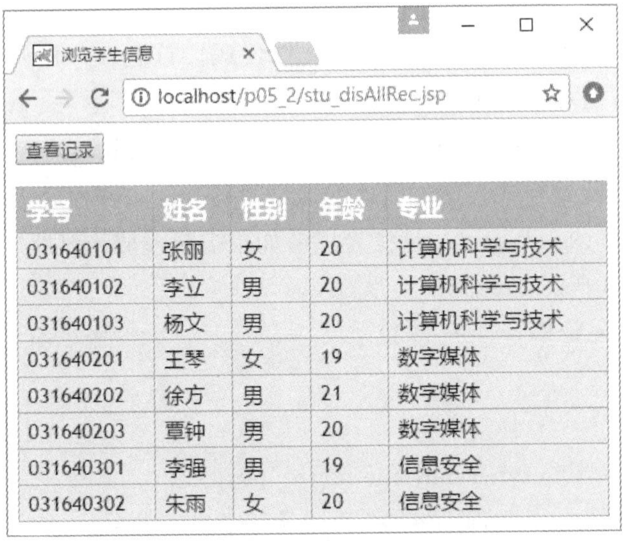

浏览学生信息

localhost/p05_2/stu_disAllRec.jsp

查看记录

学号	姓名	性别	年龄	专业
031640101	张丽	女	20	计算机科学与技术
031640102	李立	男	20	计算机科学与技术
031640103	杨文	男	20	计算机科学与技术
031640201	王琴	女	19	数字媒体
031640202	徐方	男	21	数字媒体
031640203	覃钟	男	20	数字媒体
031640301	李强	男	19	信息安全
031640302	朱雨	女	20	信息安全

图 5—6　加载 css 后效果图

四、实验思考

1. 按步骤完成实验内容，每个步骤截图保存，形成实验报告。
2. 在本实验基础上，尝试使用 JDBC 完成学生信息的增、删、改等操作。
3. 根据个人喜好，重新设计 CSS，运用到本实验中，截图保存。

实验六　JDBC 与数据库连接池

一、实验目的

1. 进一步掌握操纵数据库的基本 SQL 语句。
2. 掌握数据库连接池的基本概念。
3. 了解常用的数据库连接池技术，以及其特点。

4. 掌握 JDBC 实现数据库连接池的基本步骤和方法。

5. 掌握 JDBC 数据库连接池的 Web 应用程序开发方法。

二、基础知识

1. 数据库连接池概念。

数据库连接池就是负责管理、分配和回收数据库连接的一种技术。该技术初始化了一定数量的数据库连接,允许进程重复使用这些已经建立好的数据库连接,从而避免建立数据库连接时的性能消耗。而且,使用后的数据库连接可直接放归连接池,由连接池统一管理,用于其他连接请求,由此提高数据库访问效率和访问速度。

2. 数据库连接池技术。

主流数据库连接池技术包括:C3P0、DBCP、Tomcat Jdbc Pool、BoneCP、Proxool、Druid 等。其中 C3P0 是开源的 JDBC 连接池,实现了数据源和 JNDI 绑定,目前使用它的开源项目有 Hibernate、Spring 等;DBCP 是一个依赖 Jakarta commons－pool 对象池机制的数据库连接池,Tomcat 7. 0 之后版本使用的连接池组件就是 DBCP;Tomcat Jdbc Pool 是 Tomcat 7. 0 之前使用的连接池组件;BoneCP 是一个快速,开源的数据库连接池;Proxool 是一个 Java SQL Driver 驱动程序,提供了对选择的其他类型的驱动程序的连接池封装,可以方便移植到现存的代码中;Druid 是目前公认的较好的数据库连接池,在功能、性能、扩展性方面,都超过其他数据库连接池。

3. Druid 连接池。

Druid 是一个 JDBC 组件,提供了一个高效、功能强大、扩展性良好的数据库连接池。它包括三个部分:一个基于 Filter－Chain 模式的插件体系、DruidDataSource 高效可管理的数据库连接池,以及 SQL Parser。

同时,Druid 可对数据库的访问性能实施监控,并内置了一个功能强大的 StatFilter 插件,能够详细统计 SQL 的执行性能,这对于线上分析数据库访问性能有所帮助。另外,Druid 提供了数据库密码加密, DruidDruiver 和 DruidDataSource 都支持 PasswordCallback。而且,如果你要对 JDBC 层有编程的需求,可以通过 Druid 提供的 Filter 机制,很方便编写 JDBC 层的扩展插件。

三、实验步骤

1. 新建一个 Web 工程,在 Tomcat 中使用 Druid 连接池方式实现 Web 数据库访问。

(1)在 Eclipse 中,新建一个 Web 工程 p06_1。在网上下载 Druid 的最新版本的 jar 包,实验使用的版本是 druid－1. 1. 9. jar,将该 jar 包连同 MySQL 数据库驱动包一起导入到本工程。同时在本工程的 META－INF 目录下创建一个 context. xml 文件,用于配置数据库连接池,如图 6－1 所示。

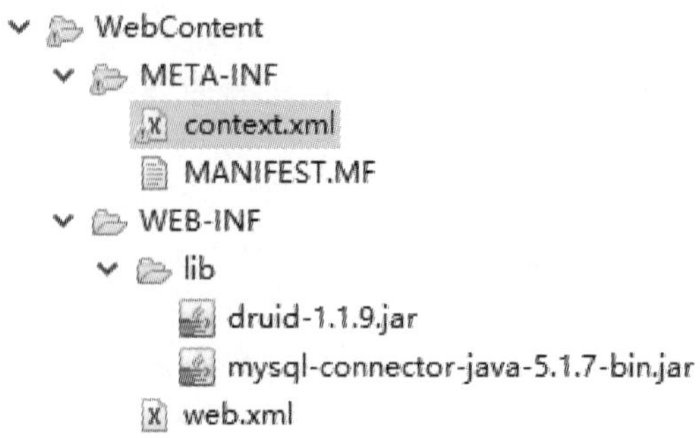

```
∨ WebContent
    ∨ META-INF
        context.xml
        MANIFEST.MF
    ∨ WEB-INF
        ∨ lib
            druid-1.1.9.jar
            mysql-connector-java-5.1.7-bin.jar
        web.xml
```

图 6—1　导入开发包

(2)打开 context. xml 文件，在 context. xml 文件中加入 JNDI 的配置信息，如下所示代码。

```
<? xml version="1. 0" encoding="UTF-8"? >
<! DOCTYPE xml>
<Context>
    <Resource
        name="jdbc/MysqlDataSource"
        factory="com. alibaba. druid. pool. DruidDataSourceFactory"
        auth="Container"
        type="javax. sql. DataSource"
        driverClassName="com. mysql. jdbc. Driver"
        url="jdbc:mysql://localhost:3306/edu_db? characterEncoding=utf-8"
        username="root"
        password="123456"
        maxActive="50"
        maxWait="10000"
        removeabandoned="true"
        removeabandonedtimeout="60"
        logabandoned="false"
        filters="stat"/>
</Context>
```

根据配置好的各项参数，连接池的工作原理如下。连接池在初始化时会创建基于 initialSize 参数的连接，当有数据库操作请求时，会从池中取出一个连接；如果当前池中正在使用的连接数等于 maxActive，则会等待一段时间，等待其他操作释放掉某一个连接，如果这个等待时间超过了 maxWait，则会报错；如果当前正在使用的连接数没有达到 maxActive，则先判断当前是否有空闲连接，如果有则直接使用空闲连接，如果没有则新建立一个连接；在连接使用完毕后，不是将其物理连接关闭，而是将其放入池中等待

其他数据库操作时复用；同时连接池内部有机制判断是否当前的总连接数少于 miniIdle，如果是，则会建立一个新的空闲连接，以保证达到最小连接数 miniIdle；如果当前连接池中某个数据库连接的空闲时间达到 timeBetweenEvictionRunsMillis 后仍未被使用，则会被关闭；当连接池中连接出现失效情况时，可通过设置 testWhileIdle 参数为 true，来保证连接池内部定时检测连接的可用性，不可用的连接会被抛弃或者重建，以尽可能保证从连接池中获得的 Connection 对象是可用的；为了保证绝对的可用性，也可以设置 testOnBorrow 参数为 true，即在获取 Connection 对象时检测其可用性，不过这样会影响性能。

(3)在工程 WebContent 目录下新建 stu_queryAllRec. jsp 文件，用于显示全体学生的记录。为了能够使用 JSTL 标签，需要下载 jstl. jar 和 standard. jar，并复制到 WEB—INF 文件夹下的 lib 目录中。stu_queryAllRec. jsp 文件部分代码如下所示。

```
<%@ page pageEncoding="UTF-8"%>
<%@ taglib uri="http://java.sun.com/jsp/jstl/core" prefix="c"%>
<%@ taglib uri="http://java.sun.com/jsp/jstl/sql" prefix="sql"%>
……
<body>
<h3>浏览 edu_stu 表记录</h3>
<sql:query var="rs" dataSource="jdbc/MysqlDataSource">
            select * from edu_stu
      </sql:query>
<c:forEach var="row" items="${rs.rows}">                ${row.stuNum}
-${row.stuName}-${row.stuSex}-${row.stuAge}-${row.stuMajor}<br>
</c:forEach>
<hr>
</body>
</html>
```

(4)查看运行效果。选择 stu_queryAllRec. jsp，点击菜单栏 Run | Run As | Run On Server，弹出对话框选择确定，可以在 Eclipse 的内置浏览器上看到运行效果。如图 6—2 所示。

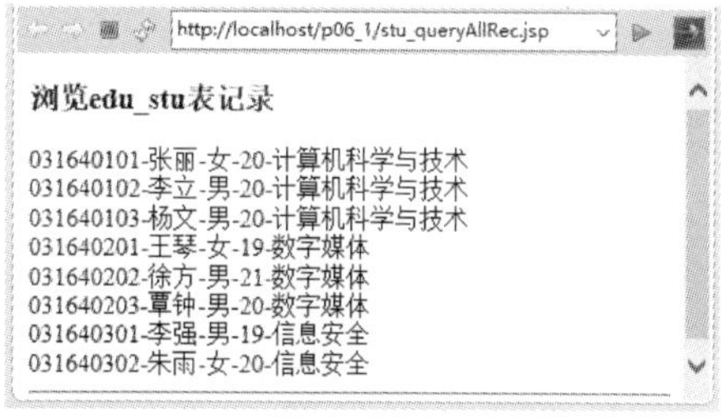

图 6—2 浏览 edu_stu 表记录

2. 新建一个 Web 工程，使用 JDBC 数据库连接类访问 Druid 连接池，并实现 Web 数据库访问。

(1)在 Eclipse 中，新建一个 Web 工程 p06_2。与上例相同，给工程导入 druid－1. 1. 9. jar 包和 MySQL 数据库驱动包。同时在工程 p06_2 的 META－INF 目录下参考上例创建 context. xml 文件。

(2)在工程中新建 DBUtil. java 类，如图 6－3 所示。

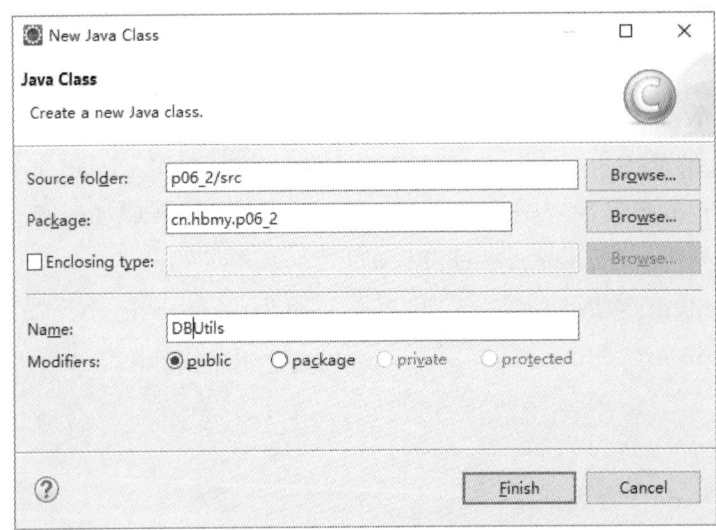

图 6－3　新建 DBUtil 数据库连接类

(3)新建的 DBUtil. java 类作为工具类，通过数据库连接池实现数据库的连接和释放，部分码如下所示。

```
package cn. hbmy. p06_2;
……
public class DBUtil {
    private static DruidDataSourcedsMySql = null;
    static{
        try {
            //1. 初始化名称查找上下文
            Context ctx = new InitialContext( );
            //2. 通过 JNDI 名称找到 DataSource
dsMySql = (DruidDataSource) ctx. lookup(MYSQL_DB_JNDINAME);
        } catch(NamingException e) {
            e. printStackTrace( );
        }
    }
```

```
    public static Connection getConnection( ) throws SQLException {
        returndsMySql. getConnection( );
    }
    public static void release(Connection conn,Statement st,ResultSet rs){
        if(rs! =null){
            try{
                rs. close( );
            }catch(Exception e) {
                e. printStackTrace( );
            }
        }
        if(st! =null){
            try{
                st. close( );
            }catch(Exception e) {
                e. printStackTrace( );
            }
        }
        if(conn! =null){
            try{
                conn. close( );
            }catch(Exception e) {
                e. printStackTrace( );
            }
        }
    }
}
```

在此数据库连接类中，先从 JNDI 容器中获取 DataSource，然后通过 DataSource 获取数据库连接。

(4)在工程中新建 stu_queryRec. jsp 文件，用于查询符合条件的学生信息，部分代码如下所示。

```
<%@ page pageEncoding="UTF-8"%>
<%@ taglib uri="http://java. sun. com/jsp/jstl/core" prefix="c"%>
……
<body>
```

```
<h3>查询符合条件的学生记录</h3>
<form action="QueryServlet" method="post">
请设置查询条件:
<select name="zdm">
<option value="stuNum">学号</option>
<option value="stuName">姓名</option>
<option value="stuSex">性别</option>
<option value="stuAge">年龄</option>
<option value="stuMajor">专业</option>
</select>
<select name="op">
<option value=">=">&gt;=</option>
<option value="<=">&lt;=</option>
<option value="=">=</option>
<option value="like">like</option>
</select>
<input type="text" name="zdz" />
<input type="submit" value="查询" />
</form>
<c:forEach var="mySqlDataMap" items="${myDataList}">
${mySqlDataMap.stuNum}-${mySqlDataMap.stuName}-
${mySqlDataMap.stuSex}-${mySqlDataMap.stuAge}-
${mySqlDataMap.stuMajor}<br>
</c:forEach>
</body>
</html>
```

在这个 JSP 页面中,通过 Form 表单将查询条件提交给了 query Servlet。因此需要定义 query Servlet 类来处理客户端请求,并把符合条件的记录通过集合对象返还给客户端 JSP 页面,最后 JSP 页面通过 JSTL 遍历显示每条记录。

(5)在工程中新建名为 Query Servlet.java 的 Servlet 文件,用于后台查询并返回查询结果,如图 6—4 所示,部分代码如下所示。

Create Servlet

Create Servlet

Specify class file destination.

Project: p06_2

Source folder: /p06_2/src　Browse...

Java package: cn.hbmy.p06_2　Browse...

Class name: QueryServlet

Superclass: javax.servlet.http.HttpServlet　Browse...

☐ Use an existing Servlet class or JSP

Class name: QueryServlet　Browse...

< Back　Next >　Finish　Cancel

图 6—4　新建 QueryServlet 类

```
//QueryServlet 代码
package cn.hbmy.p06_2;
……
@WebServlet("/QueryServlet")
……
protected void doGet(HttpServletRequest request，HttpServletResponse response)
throws ServletException，IOException {
request.setCharacterEncoding("utf8");
String zdm = request.getParameter("zdm");
String op = request.getParameter("op");
String zdz = request.getParameter("zdz");
String sql = "select * from edu_stu where " + zdm + " " + op + " " + "?";
Connection con = null;
ResultSet rs = null;
PreparedStatement pst = null;
List<Map<String, String>> dataList = new ArrayList<Map<String, String>>();
try {
con = DBUtil.getConnection();
pst = con.prepareStatement(sql);
if ("stuNum".equals(zdm) || "stuName".equals(zdm) || "stuSex".equals(zdm) || "stuMajor".equals(zdm))
pst.setString(1, zdz);
```

```
else
pst. setInt(1, Integer. valueOf(zdz));
rs = pst. executeQuery( );
while (rs. next( )) {
Map<String, String> dataMap = new LinkedHashMap<String, String>( );
dataMap. put("stuNum", rs. getString("stuNum"));
dataMap. put("stuName", rs. getString("stuName"));
dataMap. put("stuAge", String. valueOf(rs. getInt("stuAge")));
dataMap. put("stuSex", rs. getString("stuSex"));
dataMap. put("stuMajor", rs. getString("stuMajor"));
dataList. add(dataMap);
}
} catch(SQLException e) {
e. printStackTrace( );
} finally {
DBUtil. release(con, pst, rs);
}
request. setAttribute("myDataList", dataList);
request. getRequestDispatcher ( "/stu _ queryRec. jsp "). forward ( request,
response);
}
……
```

queryServlet 根据客户端请求，查询符合条件的记录并通过 List 集合对象带回给客户端 stu_queryRec. jsp。

(6)选择 stu_queryRec. jsp，点击菜单栏 Run | Run As | Run On Server，弹出对话框选择确定，可以在 Eclipse 的内置浏览器上看到运行效果。如图 6—5 所示。

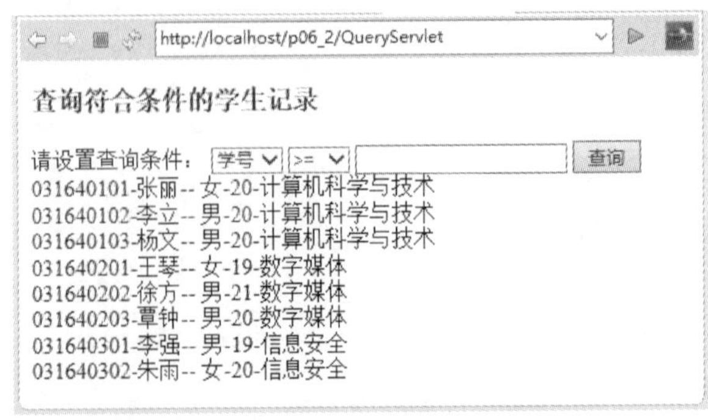

图 6—5　按条件查询记录

四、实验思考

1. 按步骤完成实验内容，每个步骤截图保存，形成实验报告。
2. 给第 2 个实验设计 CSS 文件，使运行效果更加美观。
3. 注意 Prepared Statement 的使用，并尝试将第 2 个实验中的 Prepared Statement 用 Statement 替换，并修改代码。

实验七 Hibernate 持久层开发框架

一、实验目的

1. 理解和掌握 ORM 概念。
2. 掌握 Hibernate 框架概念及工作原理。
3. 掌握 Hibernate 编程接口和实现方法。
4. 掌握 Hibernate 的关系映射。

二、基础知识

1. ORM 概念。

ORM 指的是“对象/关系映射”，是应用程序和数据库的桥梁，它可以把关系型数据包装成面向对象的模型，而完成这种从关系数据到对象模型映射的工具就是 ORM 框架。ORM 框架是面向对象程序设计语言与关系数据库发展不同步时的中间解决方案。当我们采用 ORM 框架后，应用程序就不再直接访问底层数据库，而是以面向对象的方式来操作持久化对象，然后由 ORM 框架将这些面向对象的操作转化成底层的 SQL 操作。

ORM 解决的主要问题就是对象—关系的映射。域模型和关系模型都分别建立在概念模型的基础上。域模型是面向对象的，而关系数据模型是面向关系的。一般情况下，一个持久化类和一个表对应，类的每个实例对象对应表中的一条记录。

2. Hibernate 框架。

Hibernate 是一个免费的开源 Java 包，是目前最流行的 ORM 框架之一，它是一个面向 Java 环境的 ORM 工具。它使得程序与数据库的交互变得十分容易，更加符合面向对象的设计思想，像数据库中包含普通 Java 对象一样，而不必考虑如何把它们从数据库表中读出或者写入，使得开发人员可以专注于应用程序的对象和功能，而不必关心如何在数据库中保存或查找这些对象。甚至在对 SQL 语句完全不了解的情况下，使用 Hibernate 仍然可以开发出优秀的包含数据库访问的应用程序。目前最新 Hibernate 版本是 Hibernate 5。

3. Hibernate 实现步骤。

运用 Hibernate 框架使得我们只需专注于面向对象的程序编程，而对于对象的数据库操作可交由 ORM 去完成。通常使用 Hibernate 框架的步骤如下。

(1)加载 Hibernate 类库。

(2)加载相应的数据库驱动包。

(3)创建 Hibernate 的配置文件。

(4)创建持久化类。

(5)创建对象—关系映射文件。

(6)通过 Hibernate API 编写数据库访问代码。

4. Hibernate 核心接口。

(1)Configuration 接口，用于配置和启动 Hibernate，以及创建 SessionFactory 对象。调用方法如下所示。

Configuration config=new Configuration( ). configure( );

默认情况下 Hibernate 会去 classPath 下加载 hibernate. cfg. xml 文件，如果你没有采用默认的配置文件名，那么你就需要在 configurate( )方法里面带上你的配置文件名。例如下面的方式。

File file = new File(“simpleit. xml”);

Configuration cfg = new Configuration( ). configure(file);

(2)SessionFactory 接口，用于初始化 Hibernate，充当数据存储源的代理，创建 Session 对象。SessionFactory 不是轻量级的，因为一般情况下，一个项目通常只需要一 SessionFactory，当需要操作多个数据库时，可以为每个数据库指定一个 Session Factory，其调用方法如下所示。

Configuration configiguration = new Configuration( ). configure( );

Service Registry Builder builder = new Service Registry Builder( ). apply Settings (configiguration. get Properties( ));

Service Registry registry = builder. build Service Registry( );

factory = configiguration. build Session Factory(registry);

(3)Session 接口，用于负责执行被持久化对象的 CRUD 操作，所有持久层的操作数据都缓存在 session 对象中，相当于 JDBC 中的 Connection。需要注意的是，Session 的设计是非线程安全的，即一个 Session 实例同时只可由一个线程使用。Session 实例由 SessionFactory 构建，构建方式如下。

Session session = sessionFactory. openSession( );

(4)Transaction 接口，是对 Hibernate 中的事务进行管理的接口，事务对象通过 Session 创建，例如以下语句。

Transaction ts = session. beginTransaction( );

(5) Query 和 Criteria 接口，用于对数据库及持久对象进行查询。第一步，创建 Query 实例。例如：

Query query＝session. createQuery(hql)；

第二步，用 setter 方法设置动态参数；第三步，执行查询语句，返回查询结果。返回结果既可以是 List，也可以是唯一对象。

5. Hibernate 关系映射。

Hibernate 是一个 ORM 框架，ORM 即是对象关系映射，这里的“关系”就是数据库里的数据表。也就是要建立实体对象同数据库表之间的映射，用面向对象的操作来操作数据库表数据。hibernate 在实现 ORM 功能需要用到的文件包括：实体类(*. Java)、映射文件(*. hbm. xml)和数据库配置文件(*. properties 或者 *. cfg. xml)。

Hibernate 有 7 种映射关系分别是：

(1)单向一对一关联映射(one－to－one)；

(2)单向多对一关联映射(many－to－one)；

(3)单向一对多关联映射(one－to－many)；

(4)单向多对多映射(many－to－many)；

(5)双向一对一关联映射；

(6)双向一对多关联映射；

(7)双向多对多关联映射。

以上映射关系中最重要的一种映射关系是“双向一对多关联映射”，这里以它为例说明映射关系的使用。

例如，有成绩(Score)和学生(Student)2 个实体，对应到数据库就是 2 个关系，这 2 个实体间可以建立双向的一对多关系映射。在 Student 类的映射文件 Student. hbm. xml 中加入如下标签映射，即在“一对多”关系中“一”的一方加入<key>标签。

```
<set name="scores" inverse="true">
        <key column="stuNum"/>
       <one-to-many class="cn. hibernate. Score"/>
</set>
```

在 Score 类的映射文件 Score. hbm. xml 中加入如下标签映射，即在“一对多”关系中“多”的一方加入<many－to one>标签。

```
<many-to-one name="student" class="cn. hibernate. Student" column="
stuNum"/>
```

如此建立了两个关系间“双向一对多关联映射”，其产生的效果通过实验展示。

三、实验步骤

1. 在 Eclipse 中在线安装 Hibernate 开发插件和下载开发包。

（1）进入 Hibernate Tools 的官网地址，http://tools. jboss. org/downloads/overview. html，可以看到不同版本的插件，选择下载的版本，如图 7—1 所示。

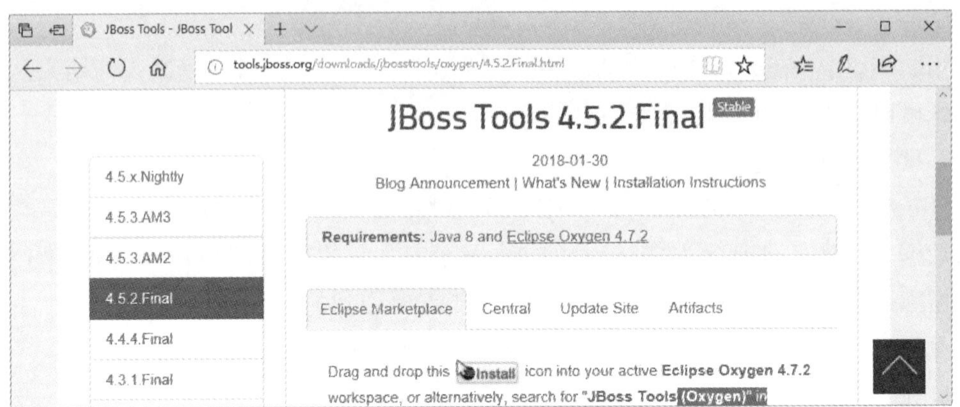

图 7—1　安装 Hibernate 插件

然后将图 7—1 中的 Install 图表拖拽到打开的 Eclipse 窗口下，将自动提示安装相应插件，按照提示完成安装过程，安装结束时提示将重启 Eclipse。

（2）下载 Hibernate 相关开发包。由于 Hibernate 版本和相应工具包的下载方式更新较快，这里不做具体方法说明，可自行网络下载。本教程所使用版本为 hibernate5. 2. 5。

2. 新建一个 Web 工程，使用 Hibernate 5 框架实现 Web 数据库访问。

（1）在 Eclipse 中，新建一个 Web 工程 p07_1。在工程中导入 Hibernate 开发包和 MySQL 数据库驱动包。导入 jar 包到 p7_1/WebContent/WEB—INF/lib 目录下，如图 7—2 所示。

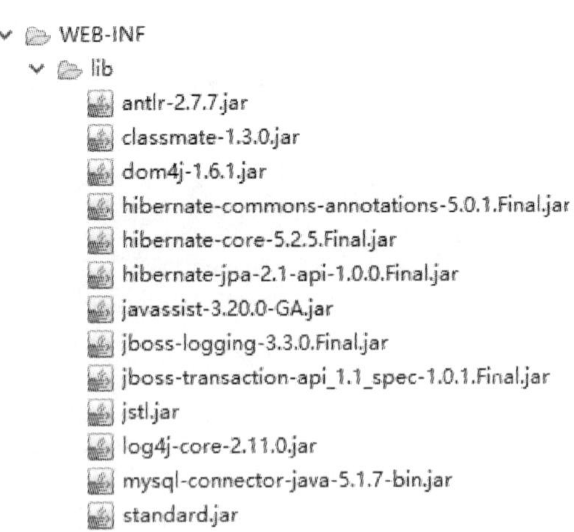

图 7—2　工程导入的工具包

（2）在工程 p07_1 的 src 目录下新建 POJO（Plain Ordinary Java Object，简单 Java 对象）类，类名为 Student，包名为 cn. hbmy. p07_1. pojo。类代码如下所示。

```
//Student. java
package cn. hbmy. p07_1. pojo;
```

```
……
@Entity
@Table(name="edu_stu")
public class Student implements java.io.Serializable {
private String stuNum;
private String stuName;
private String stuSex;
private int stuAge;
private String stuMajor;
//省略 setter 和 getter 方法
……
```

需要注意的是，通过@Entity 注解了持久化实体类，为类名和对应表名不同，因此使用@Table(name="edu_stu")注解方式将 Student 类同 edu_stu 表建立映射关系。另外，通过实现 java.io.Serializable 接口，避免了程序编译时的 java.io.InvalidClassException 错误。

(3)使用 Hibernate 插件创建对象—关系映射文件。右击 Student 类，弹出快捷菜单中选择 New | Other | Hibernate | Hibernate XML Mapping file(hbm.xml)，点击 Next，选择 Student，然后点击 Finish，可为 Student 类创建对象—关系映射文件。注意，一般都把这两个文件放在同一个目录下，如图 7—3 所示。

```
src
  cn.hbmy.p07_1.po
    Student.java
    Student.hbm.xml
```

图 7—3 新建对象—关系映射文件

生成的 Student.hbm.xml 文件部分代码如下所示。

```
<!—
……
<hibernate-mapping>
<class name="cn.hbmy.p07_1.pojo.Student" table="edu_stu">
        <id name="stuNum" type="java.lang.String">
            <column name="STUNUM" />
            <generator class="assigned" />
        </id>
        <property name="stuName" type="java.lang.String">
            <column name="STUNAME" />
        </property>
        <property name="stuSex" type="java.lang.String">
```

```
            <column name="STUSEX" />
        </property>
        <property name="stuAge" type="int">
            <column name="STUAGE" />
        </property>
        <property name="stuMajor" type="java.lang.String">
            <column name="STUMAJOR" />
        </property>
    </class>
</hibernate-mapping>
……
```

需要注意的是，文件中注明了字段名和属性名的关系，字段名 column name 都用的大写字母，这里不区分大小写。注意文件中自动注明的类名和同名的数据表间建立映射关系如下代码。

```
<class name="cn.hbmy.p07_1.pojo.Student" table="STUDENT">
```

但实际本例中的 Student 类对应的表名是 edu_stu，因此应该在此处改为实际的表名。

```
<class name="cn.hbmy.p07_1.pojo.Student" table="edu_stu">
```

(4)创建 Hibernate 配置文件。右击 src 目录，弹出快捷菜单中选择 New | Other | Hibernate | Hibernate Configuration File(cfg.xml)，点击 Next，再点击 Next，如图 7—4 所示完成 Hibernate 数据库连接配置，或者直接点击 Finish，生成配置文件后，再进行配置。

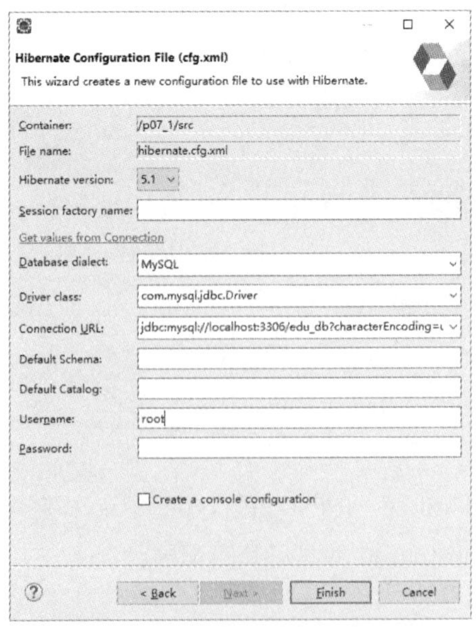

图 7—4　Hibernate 数据库连接配置

打开生成的 hibernate. cfg. xml 文件，对应的配置代码如下。

```
<! -- hibernate. cfg. xml-->
……
<session-factory>
<property name="hibernate. connection. driver_class">com. mysql. jdbc. Driver
</property>
<property name="hibernate. connection. url">
jdbc:mysql://localhost:3306/edu_db? characterEncoding=utf-8</property>
<property name="hibernate. connection. username">root</property>
<property name="hibernate. dialect">org. hibernate. dialect. MySQLDialect</
property>
<! --数据库方法配置，根据不同的方言生成符合当前数据库语法的 sql -
->
<property name="hibernate. dialect">org. hibernate. dialect. MySQL5Dialect</
property>
<! --显示 hibernate 在运行的时候执行的 sql 语句 -->
<property name="hibernate. show_sql">true</property>
<! --格式化 sql-->
<property name="hibernate. format_sql">true</property>
<! --自动建表-->
<property name="hibernate. hbm2ddl. auto">update</property>
</session-factory>
……
```

(5)在工程 p07_1 中新建 Servlet 类，用于通过 Hibernate 编程接口完成查询所有记录的操作，并将记录结果以 List 方式返还给客户端的 JSP。类名为 HibernateQueryServlet 包名为 cn. hbmy. p07_1. servlet，部分代码如下所示。

```
<! -HibernateQueryServlet. java-->
……
package cn. hbmy. p07_1. servlet;
……
@WebServlet("/HibernateQueryServlet")
……
protected void doGet(HttpServletRequest request, HttpServletResponse response)
throws ServletException, IOException {
SessionFactory sessionFactory = null;
```

```
Transaction ts = null;
Session session = null;
Configuration configuration = new Configuration( ). configure( );
configuration. addClass(Student. class);
try {
sessionFactory = configuration. buildSessionFactory( );
session = sessionFactory. openSession( );
} catch (Exception e) {
}
try {
ts = session. beginTransaction( );
Query<Student> query=session. createQuery("from Student",Student. class);
List<Student> list = query. getResultList();
ts. commit( );
session. close( );
request. getSession( ). setAttribute("dataList", list);
response. sendRedirect("stu_hQueryAllRec. jsp");
} catch (Exception e) {
ts. rollback( );
} finally {
sessionFactory. close( );
}
}
……
```

需要注意的是，代码中描述了 Hibernate 5 访问数据库的基本方法和步骤。

①加载 Hibernate 配置文件，Configuration configuration = new Configuration( ). configure( )。注意：configuration. addClass(Student. class)，这条代码是在配置中加载实体类，作用等同于在 hibernate. cfg. xml 文件中使用如下配置。

```
<mapping resource = "cn/hbmy/p07_2/pojo/Student. hbm. xml"/>
```

②创建 SessionFactory，sessionFactory = configuration. buildSessionFactory( )。

③创建 Session，session = sessionFactory. openSession( )。

④开启事务，Transaction ts = session. beginTransaction( )。

⑤执行具体数据库操作，query = session. createQuery("from Student",Student. class)。

⑥关闭 session，session. close( )。

(6)在工程 p07_1 中新建 stu_hQueryAllRec.jsp 文件,用于浏览数据。并参考实验五的方法引入样式文件 tb.css,样式文件代码见实验五。stu_hQueryAllRec.jsp 的代码如下所示。

```
<!--stu_hQueryAllRec.jsp-->
<%@ page contentType="text/html; charset=UTF-8" pageEncoding="UTF
-8"%>
<%@ taglib uri="http://java.sun.com/jsp/jstl/core" prefix="c"%>
<html>
<head>
<title>浏览学生信息</title>
<link rel="stylesheet" type="text/css" href="css/tb.css">
</head>
<body>
<form action="HibernateQueryServlet" method="post">
<input type="submit" value="查看记录" />
</form>
<table id="students">
<tr>
<th>学号</th>
<th>姓名</th>
<th>性别</th>
<th>年龄</th>
<th>专业</th>
</tr>
<c:forEach var="stu" items="${dataList}">
<tr class='alt'>
<td>${stu.stuNum}</td>
<td>${stu.stuName}</td>
<td>${stu.stuSex}</td>
<td>${stu.stuAge}</td>
<td>${stu.stuMajor}</td>
</tr>
</c:forEach>
</table>
</body>
```

</html>

程序运行效果如图 7—5 所示。

浏览学生信息

localhost/p07_1/stu_hQueryAllRec.jsp

查看记录

学号	姓名	性别	年龄	专业
031640101	张丽	女	20	计算机科学与技术
031640102	李立	男	20	计算机科学与技术
031640103	杨文	男	20	计算机科学与技术
031640201	王琴	女	19	数字媒体
031640202	徐方	男	21	数字媒体
031640203	覃钟	男	20	数字媒体
031640301	李强	男	19	信息安全
031640302	朱雨	女	20	信息安全

图 7—5　程序运行效果

2. 新建一个 Web 工程，使用 Hibernate 5 实现 Web 数据库的一对多或多对一关联访问。

(1)在 Eclipse 中，新建一个 Web 工程 p07_2。在工程中导入 Hibernate 开发包和 MySQL 数据库驱动包。导入 jar 包如上例。

(2)参考上例建立 cn. hbmy. p07_2. pojo 包和 Student 类；建立 cn. hbmy. p07_2. servlet 和 Hibernate Query Servlet 类；为工程建立配置文件 hibernate. cfg. xml；新建 stu_h Query AllRec. jsp 文件。所有相关文件代码，参考上例。

(3)建立数据表。在 MySQL 下为 edu_db 数据库中建立数据表 edu_score，具体的 DDL 如下所示。

```
Create table edu_score(
sId int(6) primary key not null auto_increment,
stuNum varchar(9) not null,
cId varchar(5) not null,
sc int(3) default 0
) default charset=utf8;
```

(4)在 cn. hbmy. p07_2. pojo 包下，新建 POJO 类、Score 类。由于 Score 类对应 edu_score 表，因此需要用注解方式进行说明，另外为了避免类的序列化错误，通过实现 java. io. Serializable 接口。具体的代码如下所示。

```
//Score. java
package cn. hbmy. p07_2. pojo;
```

```
@Entity
@Table(name="edu_score")
public class Scoreimplements java.io.Serializable {
private int sId;
private String cId;
private String stuNum;
private int sc;
private Student stu;
//省略 setter 和 getter 方法。
……
```

(5)Student 类和 Score 类之间存在着一对多关系，为了建立两者之间的一对多关系映射，需要在“一”的那一方的实体定义中，有一个私有的“多”那一方的实体对象属性，并且提供公有的 getter 和 setter 方法；同时，在“多”的那一方的实体类定义中，要有一个私有的 set 集合属性来保存“一”那一方的对象集合，并提供公有的 getter 和 setter 属性。例如，Student 类和 Score 类是一对多关系，则在前面 Student 类定义基础上，增加代码如下。

```
private Set<Score> scores=new HashSet<Score>( );
public Set<Score> getScores( ) {
return scores;
}
public void setScores(Set<Score> scores) {
this.scores = scores;
}
```

然后分别为 2 个类生成映射文件，Student.hbm.xml 文件的部分代码如下所示。

```
<!--Student.hbm.xml-->
……
<hibernate-mapping>
    <class name="cn.hbmy.p07_2.pojo.Student" table="edu_stu">
        <id name="stuNum" type="java.lang.String">
            <column name="STUNUM" />
            <generator class="assigned" />
        </id>
        <property name="stuName" type="java.lang.String">
            <column name="STUNAME" />
        </property>
```

```
            <property name="stuSex" type="java.lang.String">
                <column name="STUSEX" />
            </property>
            <property name="stuAge" type="int">
                <column name="STUAGE" />
            </property>
            <property name="stuMajor" type="java.lang.String">
                <column name="STUMAJOR" />
            </property>
            <set name="scores" table="edu_score" cascade="all">
                <key column="stuNum"></key>
                <one-to-many class="cn.hbmy.p07_2.pojo.Score"></one-
to-many>
            </set>
        </class>
    </hibernate-mapping>
```

Score.hbm.xml 文件的部分代码如下所示。

```
    <!--Score.hbm.xml-->
    ……
    <hibernate-mapping>
        <class name="cn.hbmy.p07_2.pojo.Score" table="edu_score">
            <id name="sId" type="int" access="field">
                <column name="SID" />
                <generator class="identity" />
            </id>
            <property name="cId" type="java.lang.String" access="field">
                <column name="CID" />
            </property>
            <property name="stuNum" type="java.lang.String">
                <column name="STUNUM" />
            </property>
            <property name="sc" type="int">
                <column name="SC" />
            </property>
    <many-to-one name="stu" insert="false" class="cn.hbmy.p07_2.pojo.
```

```
Student"  cascade="save-update" >
                <column name="stuNum" />
            </many-to-one>
        </class>
    </hibernate-mapping>
```

需要注意的是，这两个配置文件中<key>标签和<many-to-one>标签加入的字段(column)必须保持一致，否则会产生数据混乱。

(6)新增了类的映射文件后，在 hibernate. cfg. xml 文件中加入实体类的映射，增加代码如下。

```
<mapping resource="cn/hbmy/p07_2/pojo/Student. hbm. xml"/>
<mapping resource="cn/hbmy/p07_2/pojo/Score. hbm. xml"/>
```

(7)修改 HibernateQueryServlet 类，部分代码如下所示。

```
//HibernateQueryServlet. java
……
protected void doGet(HttpServletRequest request, HttpServletResponse response)
throws ServletException, IOException {
SessionFactory sessionFactory = null;
Transaction ts = null;
Session session = null;
Configuration configuration = new Configuration( ). configure( );
try {
session Factory = configuration. build Session Factory( );
session = session Factory. open Session( );
} catch (Exception e) {
System. out. println("生成 SessionFactory 错误!");
}
try {
Set<Score> set = new HashSet<Score>( );
ts = session. begin Transaction( );
Student stu=new Student( );
Score sc1=new Score( );
sc1. setcId("003");
sc1. setSc(71);
set. add(sc1);
stu. setScores(set);
```

```
stu. setStuNum("031640105");
session. save(stu);
Query<Student> query=session. createQuery("from Student",Student. class);
List<Student> list = query. getResultList();
ts. commit( );
session. close( );
request. getSession( ). setAttribute("dataList", list);
response. sendRedirect("stu_hQueryAllRec. jsp");
} catch (Exception e) {
ts. rollback( );
System. out. println(e. toString( ));
} finally {
sessionFactory. close( );
}
}
......
```

运行效果如图 7—6 所示。

http://localhost/p07_2/stu_hQueryAllRec.jsp

查看记录

学号	姓名	性别	年龄	专业
031640101	张丽	女	20	计算机科学与技术
031640102	李立	男	20	计算机科学与技术
031640103	杨文	男	20	计算机科学与技术
031640105			0	
031640201	王琴	女	19	数字媒体
031640202	徐方	男	21	数字媒体
031640203	覃钟	男	20	数字媒体
031640301	李强	男	19	信息安全
031640302	朱雨	女	20	信息安全

图 7—6　程序运行效果

通过 MySQL 命令访问 edu_score 表。进入命令提示符窗口，输入命令：mysql-u root-p，输入密码进入 MySQL 命令状态，访问 edu_score 表，结果图 7—7 所示。

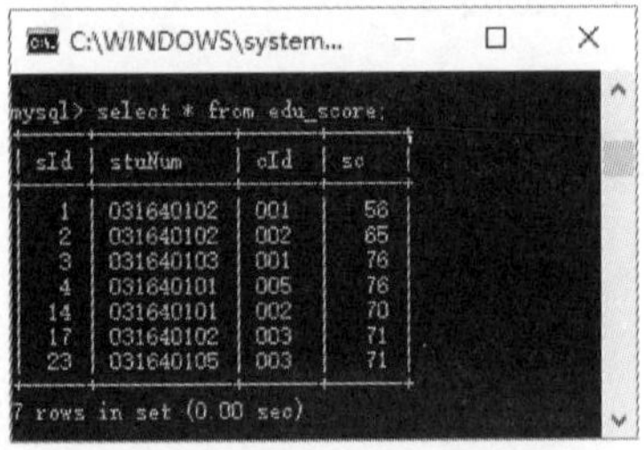

```
mysql> select * from edu_score;
| sId | stuNum    | cId | sc |
|   1 | 031640102 | 001 | 56 |
|   2 | 031640102 | 002 | 65 |
|   3 | 031640103 | 001 | 76 |
|   4 | 031640101 | 005 | 76 |
|  14 | 031640101 | 002 | 70 |
|  17 | 031640102 | 003 | 71 |
|  23 | 031640105 | 003 | 71 |
7 rows in set (0.00 sec)
```

图 7—7　edu_score 表记录

在 Hibernate Query Servlet 类中通过 session. save(stu) 方法保存了学号为031640105 的同学信息，但同时也保存了 031640105 同学的一条成绩信息于 edu_score 表中。实现这种级联操作的原因在于建立了 Student 类与 Score 类的双向的一对多映射关联，即前面对 Student. hbm. xml 和 Score. hbm. xml 文件的映射设置。

继续修改 Hibernate Query Servlet 类代码，将刚保存的 stu 对象删除，看会产生什么影响。将保存 stu 对象的代码用如下代码代替。

```
session. delete((Student)session. load(Student. class, "031640105"));
```

然后运行程序，效果如图 7—8 所示。

http://localhost/p07_2/stu_hQueryAllRec.jsp

查看记录

学号	姓名	性别	年龄	专业
031640101	张丽	女	20	计算机科学与技术
031640102	李立	男	20	计算机科学与技术
031640103	杨文	男	20	计算机科学与技术
031640201	王琴	女	19	数字媒体
031640202	徐方	男	21	数字媒体
031640203	覃钟	男	20	数字媒体
031640301	李强	男	19	信息安全
031640302	朱雨	女	20	信息安全

图 7—8 删除 031640105 后学生信息

可以看到，学号 031640105 的学生信息已被删除。然后查看 edu_score 表数据，结果如图 7—9 所示。

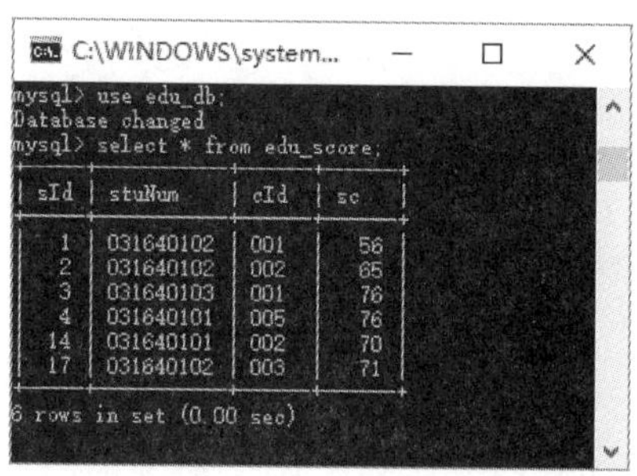

```
C:\WINDOWS\system...
mysql> use edu_db;
Database changed
mysql> select * from edu_score;
+-----+-----------+-----+----+
| sId | stuNum    | cId | sc |
+-----+-----------+-----+----+
|   1 | 031640102 | 001 | 56 |
|   2 | 031640102 | 002 | 65 |
|   3 | 031640103 | 001 | 76 |
|   4 | 031640101 | 005 | 76 |
|  14 | 031640101 | 002 | 70 |
|  17 | 031640102 | 003 | 71 |
+-----+-----------+-----+----+
6 rows in set (0.00 sec)
```

图 7—9 删除 031640105 后分数数据

发现 031640105 学号的分数数据也一并删除。

四、实验思考

1. 按步骤完成实验内容，每个步骤截图保存，形成实验报告。
2. 请尝试使用 Hibernate 分页技术实现分页查询。
3. 本实验使用了 HQL 语句进行了查询操作，请自学 HQL 的相关语法。
4. 在本实验基础上，尝试使用 Hibernate 技术完成增、删、改等其他数据库操作。
5. 总结 Hibernate 实现数据库操作的方式。

实验八　MyBatis 持久层开发框架

一、实验目的

1. 掌握 MyBatis 框架概念及工作原理。

2. 掌握 MyBatis 编程接口和实现方法。

3. 掌握 MyBatis 的 SQL 映射。

二、基础知识

1. MyBatis 概念。

MyBatis 与 Hibernate 类似，是一个持久层的框架。与 Hibernate 不同，Hibernate 可以通过 HQL 直接针对对象操作而自动生成 SQL 语句，MyBatis 是一个不完全的 ORM 框架。MyBatis 支持普通 SQL 查询、存储过程和高级映射，同时消除了几乎所有的 JDBC 代码和参数的手工设置以及结果集的检索，采用简单的 XML 或注解进行配置和原始映射。

2. MyBatis 实现步骤。

MyBatis 的实现过程可以基于 XML 配置，也可以基于注解配置，本实验主要讲解采用 XML 配置方式，而注解方式将在第四部分的实验十三中重点使用。使用 XML 配置方式的实现步骤如下。

(1)下载并导入相应 jar 包。

(2)创建数据库和数据表。

(3)为工程建立 Mybatis 的全局配置文件 Configuration. xml。

(4)根据数据表定义实体类。

(5)定义操作实体类对应数据表的 SQL 映射文件 XXXMapper. xml。

(6)在全局配置文件 Configuration. xml 中注册映射文件 XXXMapper. xml。

(7)通过 MyBatis API 编写访问数据库的代码。

3. MyBatis 核心编程接口。

(1) SqlSessionFactoryBuilder。利用其 build 方法创建 SqlSessionFactory 实例，build 方法被重载了多种形式，可利用不同资源创建 SqlSessionFactory 实例。例如，常用如下方式创建。

① public SqlSessionFactory build(InputStream inputStream)；

② public SqlSessionFactory build(Reader reader)；

这 2 个方法将 MyBatis 的核心配置文件以输入流对象传递给 MyBatis 去创建持久层框架。

(2)SqlSessoinFactory。每一个 MyBatis 以 SqlSessoinFactory 实例为核心，通过 SqlSessoinFactory 可操作 SqlSession 对象。

(3) SqlSessoin。通过 SqlSessoin 执行已映射的 SQL 语句。一般通过 SqlSessionFactory 获得该对象并执行映射的 SQL 语句。

SqlSession session= sessionFactory. openSession( );

Student stu= session. selectOne(" cn. mybatis. entity. Student. selectStuByNum ", " 0316103 ");

4. MyBatis 的 SQL 映射。

MyBatis 的 SQL 映射，是通过 XML 文件实现 SQL 语句同 Java 代码的分离，这样简化了 JDBC 编程的代码编写工作，同时提高了程序纠错时的灵活性，可直接修改 SQL 语句，而无须重新编译 Java 程序。SQL 映射文件采用 XML 格式，其顶级元素有以下几个。

(1)select，映射 sql 查询语句。

(2)insert，映射 sql 插入语句。

(3)update，映射 sql 更新语句。

(4)delete，映射 sql 删除语句。

(5)sql，就像程序中可以复用的函数一样，这个元素下放置可以被其他语句重复引用的 sql 语句。

(6)resultMap，用来描述如何从数据库查询结果集中来加载对象。

(7)cache，给定命名空间的缓存配置。

(8)cache－ref，其他命名空间缓存配置引用。

三、实验步骤

1. 安装 Eclipse 的 MyBatis 工具插件。使用 MyBatis 插件可以方便 MyBatis 框架的使用，提高编程效率。

(1)打开 Eclipse 的 Help 菜单，单击 Eclipse Marketplace，在打开对话框中查询 MyBatis 工具，如图 8－1 所示。

图 8—1　安装 MyBatis 插件

(2)找到 MyBatis Generator 1.3.6 插件,点击 Install 按钮安装即可。

2. 下载 MyBatis 开发包。MyBatis 开发包可以从官网或其他网站获取,本书所使用 jar 包为 mybatis—3.4.6.jar。

3. 新建一个 Web 工程,使用 Mybatis 3 框架实现 Web 数据库访问。

(1)在 Eclipse 中,新建一个 Web 工程 p08_1。在工程中导入 MyBatis 开发包和 MySQL 数据库驱动包,以及导入样式文件 css/tb.css。导入 jar 包到 WEB—INF/lib 目录下,如图 8—2 所示。

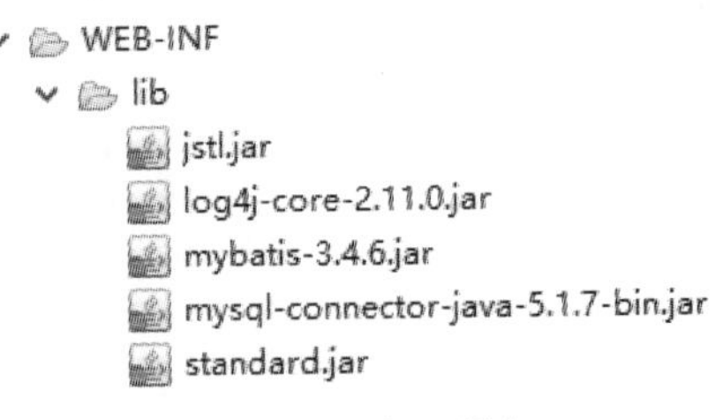

图 8—2　导入工具包

(2)创建数据库和数据表。可使用之前创建的 edu_db 数据库、edu_stu 数据表和 edu_score 数据表。

(3)利用 Eclipse 的 MyBatis 插件新建 MyBatis 配置文件。在工程中建立 resources 的资源目录,单击 resources 目录,弹出快捷菜单中选择 New | Other | MyBatis | MyBatis Generator Configuration File,确定配置文件名如图 8—3 所示,完成配置文件的新建。

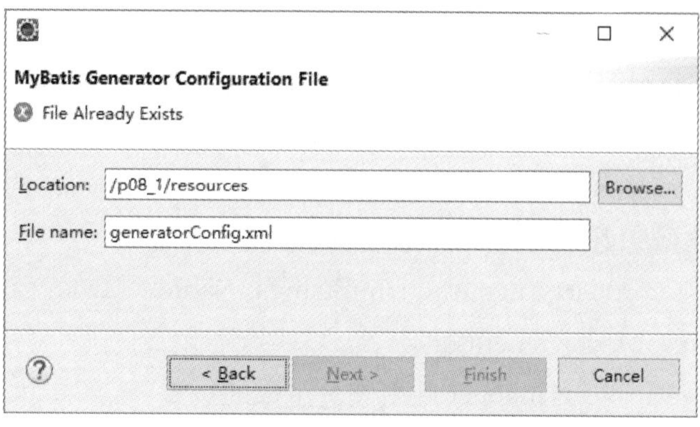

图 8—3 新建 generatorConfig. xml 配置文件

generatorConfig. xml 文件的代码如下所示。

```
<!--generatorConfig. xml-->
<? xml version="1. 0" encoding="UTF-8"? >
<! DOCTYPE generatorConfiguration PUBLIC "-//mybatis. org//DTD MyBatis Generator Configuration 1. 0//EN" "http://mybatis. org/dtd/mybatis-generator-config_1_0. dtd">
<generatorConfiguration>
    <!-- classPathEntry:数据库的 JDBC 驱动的 jar 包地址 -->
    <classPathEntry location="D:\...\mysql-connector-java-5. 1. 7-bin. jar" />
    <context id="MySql_db" targetRuntime="MyBatis3">
        <commentGenerator>
            <!--抑制警告 -->
            <property name="suppressTypeWarnings" value="true" />
            <!--是否去除自动生成的注释 true:是;false:否 -->
            <property name="suppressAllComments" value="true" />
            <!--是否生成注释代时间戳 -->
            <property name="suppressDate" value="true" />
        </commentGenerator>
        <!--数据库连接的信息:驱动类、连接地址、用户名、密码 -->
        <jdbcConnection driverClass="com. mysql. jdbc. Driver"
            connectionURL="jdbc:mysql://localhost/edu_db" userId="root"
            password="123456">
        </jdbcConnection>
    <!--生成 POJO 类的位置 -->
```

```
        <javaModelGenerator targetPackage="cn.hbmy.p08_1.pojo"
            targetProject="p08_1\src">
    <!-- enableSubPackages:是否让 schema 作为包的后缀 -->
            <property name="enableSubPackages" value="false" />
    <!-- 从数据库返回的值被清理前后的空格 -->
            <property name="trimStrings" value="true" />
        </javaModelGenerator>
    <!-- targetProject:mapper 映射文件生成的位置 -->
        <sqlMapGenerator targetPackage="cn.hbmy.p08_1.mapper"
            targetProject="p08_1\src">
    <!-- enableSubPackages:是否让 schema 作为包的后缀 -->
      <property name="enableSubPackages" value="true" />
        </sqlMapGenerator>
    <!-- targetPackage:mapper 接口生成的位置 -->
        <javaClientGenerator type="XMLMAPPER"
            targetPackage="cn.hbmy.p08_1.dao" targetProject="p08_1\
src">
            <property name="enableSubPackages" value="true" />
        </javaClientGenerator>
    <!-- 指定数据库表 -->
        <table schema="general" tableName="edu_stu" domainObjectName
="Student">
            <property name="useActualColumnNames" value="false" />
        </table>
        <table schema="general" tableName="edu_score" domainObjectName
="Score"/>
      </context>
    </generatorConfiguration>
```

（4）自动生成相关实体类、Mapper 接口类、实体的 Example 类和 Mapper.xml 映射文件。通过 Eclipse 的 MyBatis 插件可以方便地自动生成 POJO 类文件，以及 Mapper 映射文件，但前提是在上一步中设置 generatorConfig.xml 配置文件的代码必须正确，特别是对于 Table 的设置部分。当设置好 generatorConfig.xml 配置文件后，右击文件名，弹出快捷菜单中选择 Run As | My Batis Generator，将自动按配置文件生成相应文件，文件结构如图 8—4 所示。注意，自动生成相关文件的菜单命令在不同版本的 Eclipse 中的使用可能不同。

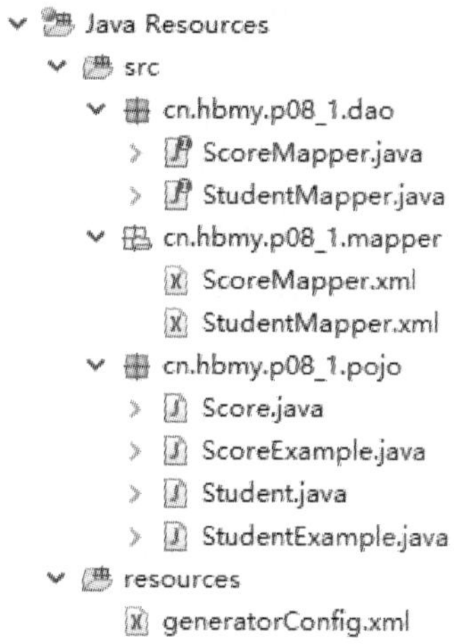

```
v Java Resources
  v src
    v cn.hbmy.p08_1.dao
      > ScoreMapper.java
      > StudentMapper.java
    v cn.hbmy.p08_1.mapper
        ScoreMapper.xml
        StudentMapper.xml
    v cn.hbmy.p08_1.pojo
      > Score.java
      > ScoreExample.java
      > Student.java
      > StudentExample.java
  v resources
      generatorConfig.xml
```

图 8—4 generatorConfig. xml 自动生成文件

Student. java 和 Score. java 类是根据数据库 edu_db 的 edu_stu 和 edu_score2 个数据表自动生成的对应 POJO 类，请注意其定义的成员变量名。StudentMapper. java 和 ScoreMapper. java 定义的是 2 个接口，是针对 Student 和 Score 定义的 Mapper 接口方法。类似于 Hibernate 的 DAO 接口方式，不同的是这里不需要实现该 DAO 接口，只需要通过 StudentMapper. xml 和 ScoreMapper. xml 文件做好方法和 SQL 的映射关系。

请读者尝试将指定数据库表的代码修改成如下代码，然后重新自动生成相关文件，并比较两种情况下的异同。

```
<table schema="general" tableName="edu_stu"
domainObjectName="Student" enableCountByExample="false"
enableUpdateByExample="false" enableDeleteByExample="false"
enableSelectByExample="false" selectByExampleQueryId="false">
<property name="useActualColumnNames" value="true" />
</table>
```

(5)在 resources 目录下新建 jdbc. properties 和 Mybatis－Config. xml 配置文件，jdbc. properties 文件代码如下所示。

```
<! --jdbc. properties-->
driver=com. mysql. jdbc. Driver
url=jdbc:mysql://localhost:3306/edu_db? characterEncoding=utf8
username=root
password=123456
driver. encode=utf-8
poolMaximumActiveConnections=15
poolMaximumIdleConnections=10
poolMaximumCheckoutTime=20000
```

创建的 Mybatis-Config. xml 文件代码如下所示。

```
<! --Mybatis-Config. xml-->
<? xml version="1. 0" encoding="UTF-8"? >
<! DOCTYPE configuration PUBLIC "-//mybatis. org//DTD Config 3. 0//EN"
  "http://mybatis. org/dtd/mybatis-3-config. dtd">
<configuration>
<! --引入 JDBC 配置文件-->
<properties resource="db. properties"> </properties>
<! --对事务的管理和连接池的配置,可以配置多个环境,默认使用 default 环境
-->
<environments default="development">
<! --id 为 development 的环境对事务的管理和连接池的配置-->
<environment id="development">
<! --事务配置,使用 jdbc 类型的事务  -->
<transactionManager type="JDBC" />
<! --数据源的配置,使用库链接池类型 -->
<dataSource type="POOLED">
<property name="driver" value="${driver}" />
<property name="url" value="${url}" />
<property name="username" value="${name}" />
<property name="password" value="${password}" />
<property name = " poolMaximumActiveConnections " value = "
${poolMaximumActiveConnections}" />
<property name = " poolMaximumCheckoutTime " value = "
${poolMaximumCheckoutTime}" />
<property name = " poolMaximumIdleConnections " value = "
${poolMaximumIdleConnections}" />
<property name="driver. encode" value="${driver. encode}" />
</environment>
</environments>
<mappers>
<! --映射路径 -->
<mapper resource="cn/hbmy/p08_1/mapper/StudentMapper. xml" />
<mapper resource="cn/hbmy/p08_1/mapper/ScoreMapper. xml" />
</mappers>
```

</configuration>

需要注意的是，Mybatis－Config. xml 文件和 GeneratorConfig. xml 文件的功能区别，前者用于加载数据库驱动和映射，后者为了简化编程自动生成相关文件。且需要注意 2 个文件的引入 dtd 的不同。

通过 GeneratorConfig. xml 可以快速自动生成相关 POJO 类、Mapper 接口类和对于 mapper. xml 映射文件，简化了编程，但必须能够理解这些文件的相互关联和作用，必须能够掌握自定义这些文件的方法。

(6)程序运行效果如图 8－5 所示。

浏览学生信息

localhost/p08_1/stu_MybatisQueryAllRec.jsp

查看记录

学号	姓名	性别	年龄	专业
031640101	张丽	女	20	计算机科学与技术
031640102	李立	男	20	计算机科学与技术
031640103	杨文	男	20	计算机科学与技术
031640201	王琴	女	19	数字媒体
031640202	徐方	男	21	数字媒体
031640203	覃钟	男	20	数字媒体
031640301	李强	男	19	信息安全
031640302	朱雨	女	20	信息安全

图 8－5　程序运行效果

4. 新建一个 Web 工程，使用 Mybatis 3 实现实体间的关联映射。

(1)在 Eclipse 中，参考上例方法新建工程 p08_2，工程文件结构如图 8－6 所示。注意文件所在包名的变化。同时导入相关开发 jar 包和 css 文件。

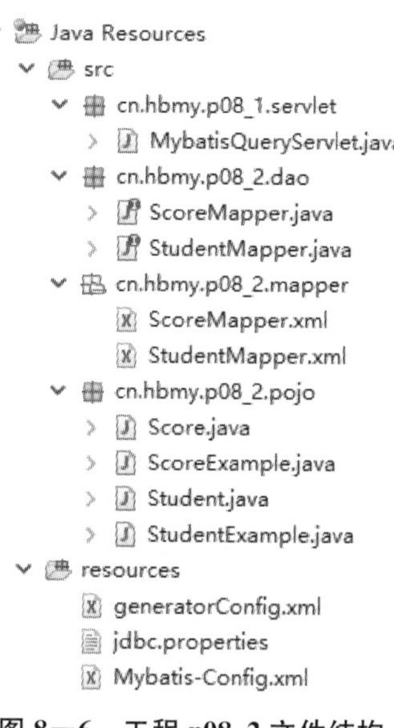

图 8－6　工程 p08_2 文件结构

(2)通过 generatorConfig. xml 自动生成相关的实体类、Mapper 接口类、Mapper 映射文件和实体的 Example 类。其中,实体类 Student 和实体类 Score 之间存在一对多关系,可以建立一对多关系映射。在 Student 类中,增加成员变量 scores,具体代码如下所示。

```
//Student. java
……
privateList< Score > scores;
//省略相应 setter 和 getter 方法
……
```

在 Score 类中,增加成员变量 stu,具体代码如下。

```
//Score. java
……
private Student stu;
//省略相应 setter 和 getter 方法
……
```

(3)修改 StudentMapper. xml 文件,添加关联映射的内容。代码如下所示。

```
<! --StudentMapper. xml-->
……
<resultMap id="BaseResultMap" type="cn. hbmy. p08_2. pojo. Student">
……
<collection property="scores" column="stunum" fetchType="lazy"
javaType="ArrayList" ofType="cn. hbmy. p08_2. pojo. Score"
select="cn. hbmy. p08_2. dao. ScoreMapper. selectScoreByStuNum">
<id property="stunum" column="stunum" />
<result property="cid" column="cid" />
<result property="sc" column="sc" />
</collection>
</resultMap>
……
```

代码中 property 表示返回类型 Student 的属性 scores;column 表示将 stunum 作为之后进行查询的参数;fetchType="lazy"表示懒加载, 即加载一个实体时,懒加载的属性不会马上从数据库中加载;javaType 表示属性对应的类型;ofType 表示集合当中的类型;select 表示的对应 cn. hbmy. p08_2. dao. ScoreMapper 接口类定义的 selectScoreByStuId 方法。

(4)修改 ScoreMapper. xml 文件,添加 SQL 语句映射。注意: mapper 接口名必须和相应 mapper. xml 的 namespace 一致,方法名和参数名及返回类型也要与 mapper. xml 的配置

一致。具体代码如下。

```
<!--ScoreMapper.xml-->
……
  <select id="selectScoreByStuNum" parameterType="String" resultType =
"cn.hbmy.p08_2.pojo.Score">
    select * from edu_score where stunum=#{stunum}
  </select>
……
```

(5)给StudentMapper接口类新增一个selectStudentByStuNum方法,添加的代码如下。

```
//StudentMapper.java
……
Student selectStudentByStuNum(String stunum);
……
```

(6)新建MybatisQueryScoreServlet类,实现按学号查询分数的功能。为工程新建Servlet,部分类代码如下所示。

```
//MybatisQueryScoreServlet.java
……
protected void doGet(HttpServletRequest request, HttpServletResponse response)
throws ServletException, IOException {
String resource = "Mybatis-Config.xml";
Student stu= new Student( );
List<Score> list=new ArrayList<Score>( );
InputStream inputStream = Resources.getResourceAsStream(resource);
SqlSessionFactory sqlSessionFactory =
new SqlSessionFactoryBuilder( ).build(inputStream);
SqlSession session = sqlSessionFactory.openSession( );
String stunum=request.getParameter("stunum");
try {
StudentMapper studentMapper =
(StudentMapper) session.getMapper(StudentMapper.class);
stu = studentMapper.selectStudentByStuNum(stunum);
session.commit( );
list=stu.getScores( );
request.getSession( ).setAttribute("dataList", list);
```

```
request. getSession( ). setAttribute("stuName", stu. getStuname( ));
request. getSession( ). setAttribute("flag", "score");
response. sendRedirect("stu_MybatisQueryRec. jsp");
} catch (Exception e) {
System. out. println(e. toString( ));
} finally {
session. close( );
}
}
......
```

(5)新建 stu_MybatisQueryRec. jsp 文件,部分关键代码如下所示。

```
<! --stu_MybatisQueryRec. jsp-->
......
<body>
<form action="MybatisQueryServlet" method="post">
点击查看所有学生记录<input type="submit" value="查看记录" />
</form>
<form action="MybatisQueryScoreServlet" method="post">
输入需要查询分数的学号:<input type="text" name="stunum"/>
<input type="submit" value="查看分数" />
</form>
<c:if test="${flag=='all'}">
<table id="students">
<tr>
<th>学号</th>
<th>姓名</th>
<th>性别</th>
<th>年龄</th>
<th>专业</th>
</tr>
<c:forEach var="stu" items="${dataList}">
<tr class='alt'>
<td>${stu. stunum}</td>
<td>${stu. stuname}</td>
<td>${stu. stusex}</td>
```

```
<td>${stu.stuage}</td>
<td>${stu.stumajor}</td>
<tr>
</c:forEach>
</table>
</c:if>
<c:if test="${flag=='score'}">
<table id="students">
<tr>
<th>学号</th>
<th>姓名</th>
<th>课程号</th>
<th>分数</th>
</tr>
<c:forEach var="score" items="${dataList}">
<tr class='alt'>
<td>${score.stunum}</td>
<td>${stuName}</td>
<td>${score.cid}</td>
<td>${score.sc}</td>
<tr>
</c:forEach>
</table>
</c:if>
</body>
</html>
```

(6)程序运行效果如图 8—7 所示。

图 8—7　程序运行效果

四、实验思考

1. 按步骤完成实验内容，每个步骤截图保存，形成实验报告。
2. 实验过程中，Sql 语句和 Java 代码完全分离，程序是如何实现数据库操作的。
3. 在本次实验基础上，练习 MyBatis 下的其他数据操作，如增、删、改等操作。
4. 思考并体会 MyBatis 的数据访问逻辑与 JDBC 的数据访问逻辑的差别。

第三部分　MVC 模式开发

实验九　MVC 与 DAO 开发模式

一、实验目的

1. 理解 DAO 设计模式的概念和设计思想。

2. 掌握 DAO 设计模式开发应用程序的基本方法。

3. 理解 MVC 和 DAO 开发模式，并熟练掌握 JSP＋Java Bean＋Servlet 的 DAO 开发模式。

4. 实现 DAO 模式的 Java Web 应用开发。

二、基础知识

1. MVC 模式。

MVC 全名是 Model View Controller，是模型（model）、视图（view）和控制器（controller）的缩写，是一种软件设计模式。强调软件设计的分层，视图层着重于设计与用户交互的个性定制界面，对应 JSP；模型层强调业务逻辑和数据逻辑设计，对应 JavaBean；控制层专注于将视图层的请求数据发送给模型层处理，对应 Servlet。

2. DAO 设计模式。

DAO 即数据访问对象（Data Access Object）。DAO 设计模式的具体做法就是将所有对数据源的访问操作抽象封装在一组公共 API 接口中，其目的就是将底层数据访问逻辑和中间层业务逻辑分离，当需要修改业务流程时，则只需调用业务接口，而无须修改对底层数据的访问。

3. DAO 模式的组成。

DAO 模式的根本在于建立数据表同对象的映射关系，并将所有对象操作原子化，在服务器端进行封装，而客户端只需要调用对象操作接口，接口的实现由服务器端完成。具体组成如下。

（1）一个专门负责数据库连接的操作类。

（2）对应数据表抽象的 VO 类（Value Object）。

（3）对应数据对象操作所抽象的 DAO 接口。

(4)对应 DAO 接口方法的具体实现类。

(5)DAO 操作模板,定义了对 VO 对象的添加、修改、删除及查询等方法。

(6)获取 DAO 实例化对象的工厂类。

三、实验步骤

1. 新建一个 Web 工程,使用 DAO 开发模式实现数据库访问的 Web 应用。

(1)在 Eclipse 中,新建一个 Web 工程 p09_1。导入相关开发 jar 包到本工程的 WEB-INF 目录下的 lib 文件中,包括 Druid 连接池、MySQL 数据库驱动、JSTL 标签库等。引入之前实验所用的 css/tb. css 样式表文件,同时在本工程的 META-INF 目录下创建一个 context. xml 文件,并完成连接池配置。具体方法参考实验六。

(2)新建一个专门负责数据库连接的操作类。具体方法和代码设计参考实验六中 DBUtil 类的设计。

(3)定义需要操作的 VO 类。这里需要操作的 VO 类,是根据 edu_stu 表的数据抽象出来的 Student 类。在工程 p09_1 中新增 Student 类,具体代码如下所示。

```
//student. java
package cn. hbmy. p09_1. vo;
public class Student {
private String stuNum;
private String stuName;
private String stuSex;
private int stuAge;
private String stuMajor;
//省略 setter 和 getter 方法
……
```

(4)抽象 DAO 接口。在工程中新建 StuDao 接口,用于封装对 Student 对象的所有原子操作的方法。右击 src 目录,选择 New | Interface,打开对话框设置新建接口,如图 9-1 所示。

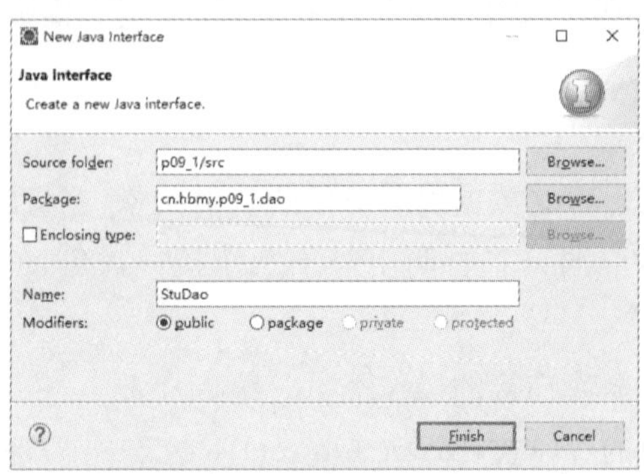

图 9-1 新建 StuDao 接口

StuDao 接口完整代码如下所示。

```
package cn. hbmy. p09_1. dao;
import java. sql. SQLException;
import java. util. List;
import cn. hbmy. p09_1. vo. * ;
public interface StuDao {
public List<Student> findByX(String param) throws SQLException;
}
```

本例中对 Student 对象只完成了按条件查询操作，故这里只声明了一个接口方法。注意接口方法只声明，不实现。

(5)定义 DAO 接口方法的具体实现类。在工程中新增一个 DAO 接口方法的具体实现类 StuImpl，右击 src 目录，选择 New | Class，新建一个名为 StuImpl 的类，如图 9—2 所示。

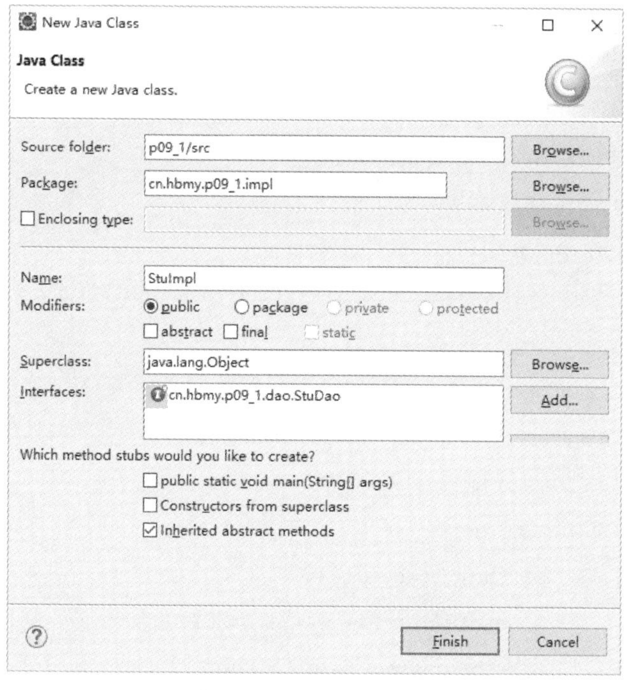

图 9—2 新增 StuDao 接口方法的实现类 StuImpl

注意，在图 9—2 中，要添加 StuImpl 类的父接口 StuDao，并在 Inherited abstract methods 处打勾，表示继承其抽象方法。StuImpl 类的主要代码如下所示。

```
//StuImpl. java
……
public class StuImpl implements StuDao {
public List<Student> findByX(String param) throws SQLException {
        List<Student>st=new ArrayList<Student>( );
```

```
            String sql="select stuNum,stuName,stuSex,stuAge,stuMajor from edu
_stu where "+ param;
            StuDaoOperateTemplate stuDtp =new StuDaoOperateTemplate( );
            st=stuDtp. queryBySql(sql);
    return st;
    }
    }
    ……
```

(6)定义 DAO 模板类。在工程中新建 StuDaoOperateTemplate 类,用于实现具体的数据库的增、删、改、查操作。StuDaoOperateTemplate 类代码如下所示。

```
    //StuDaoOperateTemplate. java
    ……
    public class StuDaoOperateTemplate {
    public List<Student> queryBySql(String sql) throws SQLException {
    Connection con=null;
            PreparedStatement pst=null;
            ResultSet rs=null;
            Student p=null;
            List<Student> st=new ArrayList<Student>( );
            try {
                con=DBUtil. getConnection( );
                pst=con. prepareStatement(sql);
                rs=pst. executeQuery( );
                while(rs. next( )){
                    p=new Student( );
                    p. setStuNum(rs. getString(1));
                    p. setStuName(rs. getString(2));
                    p. setStuSex(rs. getString(3));
                    p. setStuAge(rs. getInt(4));
                    p. setStuMajor(rs. getString(5));
                    st. add(p);
                    }
                } catch(SQLException e) {
                // TODO: handle exception
                throw new SQLException("查询所有数据异常");
```

```
            }finally {
                DBUtil.release(con, pst, rs);
            }
return st;
}
}
……
```

在 DAO 模板类中定义了对于数据库操作的具体实现方法，通常就是增、删、改、查等方法。这属于底层的数据操作，基本不会修改，当应用程序的业务逻辑需要修改时，只需要修改 DAO 接口、DAO 接口的实现类等等。

(7)为工程新建 DAO 工厂类，通过 DAO 工厂类可实例化 DAO 接口具体实现类。具体新建 DAOFactory 类，代码如下所示。

```
//DAOFactory.java
……
public class DAOFactory {
public static StuDao getStuImpl( ){
return new StuImpl( );
}
}
……
```

本例只抽象了一个 VO 类即 Student 类，也只抽象了针对这一个 VO 类的接口方法、以及实现类，所以在这个工厂类中，只定义了获取 StuImpl 类对象的方法。

到此，本工程的 DAO 设计部分已经完成，所涉及的类的具体包路径如图 9－3 所示。

```
Java Resources
  src
    cn.hbmy.p09_1.dao
      StuDao.java
    cn.hbmy.p09_1.factory
      DAOFactory.java
    cn.hbmy.p09_1.impl
      StuImpl.java
    cn.hbmy.p09_1.template
      StuDaoOperateTemplate.java
    cn.hbmy.p09_1.utils
      DBUtils.java
    cn.hbmy.p09_1.vo
      Student.java
```

图 9－3　工程类文件的包路径图

(8)DAO 模式只是解决业务逻辑和数据处理逻辑的分层，本工程中，还需要定义一个 Servlet 类来负责 DAO 同客户端的交互。在工程中新建 Servlet 类 QueryServlet，如图 9—4 所示。

图 9—4　新建 QueryServlet 类

其关键代码如下所示。

```
//QueryServlet.java
……
@WebServlet("/QueryServlet")
……
protected void doGet(HttpServletRequest request, HttpServletResponse response)
throws ServletException, IOException {
request.setCharacterEncoding("utf8");
String zdm = request.getParameter("zdm");
String op = request.getParameter("op");
String zdz = request.getParameter("zdz");
String condition = "";
if("stuAge".equals(zdm))
condition = zdm+op+zdz;
else
condition = zdm+" "+op+" '"+zdz+"'";
StuImpl stuImpl=(StuImpl) DAOFactory.getStuImpl( );
List<Student> dataList=new ArrayList<Student>( );
try {
dataList = stuImpl.findByX(condition);
```

```
} catch (SQLException e) {
e.printStackTrace();
}
request.setAttribute("myDataList", dataList);
request.getRequestDispatcher("/stu_queryRec.jsp").forward(request, response);
}
......
```

(9)新建 JSP 文件,用于访问查询结果集。在工程中新建 JSP 文件 stu_queryRec.jsp,其部分代码如下所示。

```
<!--stu_queryRec.jsp-->
......
<%@ taglib uri="http://java.sun.com/jsp/jstl/core" prefix="c"%>
......
<link rel="stylesheet" type="text/css" href="css/tb.css">
......
<form action="QueryServlet" method="post">
请设置查询条件:
<select name="zdm">
<option value="stuNum">学号</option>
<option value="stuName">姓名</option>
<option value="stuSex">性别</option>
<option value="stuAge">年龄</option>
<option value="stuMajor">专业</option>
</select> <select name="op">
<option value=">=">&gt;=</option>
<option value="<=">&lt;=</option>
<option value="=">=</option>
</select>
<input type="text" name="zdz" /> <input type="submit" value="查询" />
</form>
<table id="students">
......
<c:forEach var="stu" items="${myDataList}">
<tr class='alt'>
```

```
<td>${stu.stuNum}</td>
<td>${stu.stuName}</td>
<td>${stu.stuSex}</td>
<td>${stu.stuAge}</td>
<td>${stu.stuMajor}</td>
<tr>
</c:forEach>
</table>
……
```

Web 应用程序运行效果如图 9—5 所示。

查询记录

localhost/p09_1/QueryServlet

查询符合条件的学生记录

请设置查询条件：学号 >= 查询

学号	姓名	性别	年龄	专业
031640101	张丽	女	20	计算机科学与技术
031640102	李立	男	20	计算机科学与技术
031640103	杨文	男	20	计算机科学与技术
031640201	王琴	女	19	数字媒体
031640202	徐方	男	21	数字媒体
031640203	覃钟	男	20	数字媒体
031640301	李强	男	19	信息安全
031640302	朱雨	女	20	信息安全

图 9—5　工程 p09_1 运行效果图

四、实验思考

1. 按步骤完成实验内容，每个步骤截图保存，形成实验报告。

2. 比较本例和实验六中的 Query Servlet 类的定义有什么异同，体会 DAO 模式设计思想。

3. 本实验运行时，如果查询条件的文本框输入错误条件，比如输入类似“学号>JAVA”，观察会发生什么情况？如果希望对输入错误条件给出错误提示，该如何实现？

4. 尝试为 Student 对象定义更多其他方法，比如定义一个插入方法，则相应类和接口将如何修改。

实验十　Struts 开发框架

一、实验目的

1. 掌握 Struts 2 框架概念及工作原理。

2. 掌握 Struts 2 编程接口和实现方法。

3. 实现用户验证登录的数据库 Web 应用。

4. 理解和掌握 Struts2 国际化概念和实现方法。

二、基础知识

1. Struts 2 框架。

Struts 2 框架是一个基于 MVC 设计模式的 Web 应用框架，与 Sturts1 不同，Struts2 采用的是 Web Work 的内核，使用拦截器机制处理用户请求，运用 OGNL 表达式和 Struts2 标签表达应用程序数据，同时引入注解和约定使得框架更易于使用。

2. Struts 2 工作原理。

Struts 2 的工作原理如图 10—1 所示。

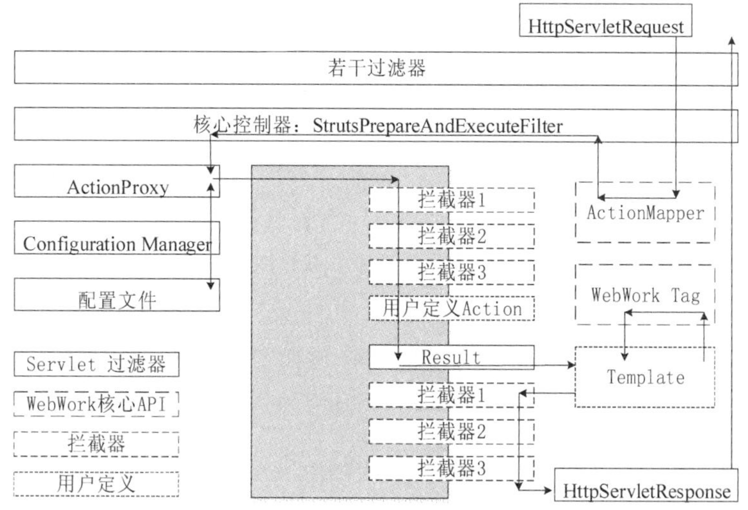

图 10—1　Struts2 工作原理图

工作流程如下。

(1)客户端通过浏览器发送基于 Http 协议的请求。

(2)Web 容器将请求传递给一系列过滤器，作简单处理后交给核心控制器 StrutsPrepareAndExecuteFilter。

(3)StrutsPrepareAndExecuteFilter 通过询问 ActionMapper 确定是否需要调用某个 Action 来处理这个请求。

(4)如果 ActionMapper 确定需要调用某个 Action,StrutsPrepareAndExecuteFilter 将请求的处理交给 ActionProxy。

(5)ActionProxy 通过 Configuration Manager 询问框架的配置文件,找到需要调用的 Action 类。

(6)ActionProxy 创建一个 ActionInvocation 的实例。

(7)ActionInvocation 实例使用命名模式来调用,在调用 Action 的过程前后,涉及相关拦截器(Intercepter)的调用。

(8)一旦 Action 执行完毕,ActionInvocation 负责根据 struts. xml 中的配置找到对应的返回结果。

3. Struts2 国际化。

国际化简称 i18n,其来源于英文单词 internationalization 的首位字母,以及取中间字母个数 18 作为名字。其根本作用是使得开发的软件支持多国语言环境。Struts2 整合了 Java 中实现的国际化功能,以方便开发者使用。其国际化包括以下内容。

(1)JSP 页面的国际化。

(2)Action 的国际化。

(3)转换错误信息的国际化。

(4)校验错误信息的国际化。

三、实验步骤

1. 从官网或其他网站获取 Struts2 开发包。

(1)访问 Apache Struts 官网,选择下载 Struts 版本,实验中使用的是 Struts 2. 5. 16 版本,下载时建议选择 Full Distribution 选项,该版本包含 Struts2 的核心库、源代码、文档和实例等。

(2)将下载文件解压缩,打开解压后的文件夹,可看到 lib 文件夹,里面是所有 Struts2 的开发 jar 包。其中 Struts2 的核心 jar 包如图 10—2 所示。

commons-fileupload-1.3.3.jar
commons-io-2.5.jar
commons-lang3-3.6.jar
freemarker-2.3.26-incubating.jar
javassist-3.20.0-GA.jar
log4j-api-2.10.0.jar
ognl-3.1.15.jar
struts2-core-2.5.16.jar

图 10—2　Struts2 的核心 jar 包

2. 新建一个 Web 工程,使用 Struts2 框架+DAO 模式实现 Web 登陆访问。

(1)在 Eclipse 中,新建一个 Web 工程 p10_1。在工程中导入 Struts2 核心开发包和 MySQL 数据库驱动包,数据库连接池 Druid,以及其他相关 jar 包,导入 jar 包到 p10_1/

WebContent/WEB－INF/lib 目录下，如图 10－3 所示。

```
lib
  commons-fileupload-1.3.3.jar
  commons-io-2.5.jar
  commons-lang3-3.6.jar
  druid-1.1.9.jar
  freemarker-2.3.26-incubating.jar
  javassist-3.20.0-GA.jar
  jstl.jar
  log4j-api-2.10.0.jar
  mysql-connector-java-5.1.7-bin.jar
  ognl-3.1.15.jar
  standard.jar
  struts2-core-2.5.16.jar
```

图 10－3　工程导入 jar 包

另外导入样式文件 css/tb. css，同时在本工程的 META－INF 目录下创建一个 context. xml 文件，并参考实验九，完成连接池配置。

(2)为工程添加数据库连接类。新建 cn. hbmy. p10_1. utils 包，在包中新建数据库连接类 DBUtil，具体代码参考实验九中 DBUtil 类。

(3)在数据库 edu_db 中创建数据表 edu_user。进入命令提示符窗口，输入命令 mysql －u root － p，然后输入密码，进入 mysql 命令行模式。输入命令 use edu_edu；访问数据库，最后输入如下命令创建数据表 edu_user。

```
CREATE TABLE edu_user(
userId int(6) primary key not  null  auto_increment,
userNum varchar(9) not null,
userPwd varchar(8) not null,
userType varchar(6) not null
) default charset=utf8;
```

(3)根据数据表，创建 Domain。新建 cn. hbmy. p10_1. domain 包，再新建 3 个 pojo 类 User、Student 和 Score。User 类部分代码如下所示。

```
//User. java
package cn. hbmy. p10_1. domain;
public class User {
private int userId;
private String userName;
private String userPwd;
private String userType;
//省略 setter 和 getter 方法
……
```

Student 类部分代码如下所示。

```
//Student. java
package cn. hbmy. p10_1. domain;
public class Student {
private String stuNum;
private String stuName;
private String stuSex;
private Integer stuAge;
private String stuMajor;
//省略 setter 和 getter 方法
……
```

Score 类部分代码如下所示。

```
//Score. java
package cn. hbmy. p10_1. domain;
public class Score {
private Integer sId;
private String stuNum;
private String cId;
private Integer sc;
//省略 setter 和 getter 方法
……
```

(4)针对每个数据对象封装 DAO 操作接口。为 User 类、Student 类封装 DAO 操作接口，User 类的操作包括 queryByUserName(按用户名查询用户)和 insertUser(插入用户信息)，Student 类的操作包括 queryAllStu(查询所有学生记录)、queryStuByStuNum(按学号查询学生记录)和 queryAllStuByCondition(按条件查询学生记录)。在工程中新建 con. hbmy. p10_1. dao 包，在包中新建 UserDao 接口和 StudentDao 接口，UserDao 代码如下所示。

```
//UserDao. java
package cn. hbmy. p10_1. dao;
import cn. hbmy. p10_1. domain. User;
public interface UserDao {
public User queryByUserName(String userName);
public boolean insertUser(String userName,String userPwd,String userType);
}
```

StudentDao 代码如下所示。

```
//StudentDao.java
package cn.hbmy.p10_1.dao;
import java.util.List;
import cn.hbmy.p10_1.domain.Student;
public interface StudentDao {
public List<Student> queryAllStu();
public List<Student> queryAllStuByCondition(String condition);
public Student queryStuByStuNum(String stuNum);
}
```

注意:DAO类中通常封装了数据对象的数据库操作方法,通常都是原子操作,即最基础的操作,通常通过若干基础操作可以构成业务逻辑,且返回的是对象类型。

(5)建立DAO实现类。在工程中新建cn.hbmy.p10_1.dao.impl包,在包中新建DAO接口的实现类UserDaoImpl和StudentDaoImpl。该类用户实现了DAO接口方法,是数据库的底层操作。UserDaoImpl类的部分代码如下所示。

```
//UserDaoImpl.java
……
public class UserDaoImpl implements UserDao {
Connection conn = null;
PreparedStatement pst = null;
ResultSet rs = null;
public User queryByUserName(String userName) {
String sql = "select * from edu_user where userName=?";
try {
conn = DBUtil.getConnection();
pst = conn.prepareStatement(sql);
pst.setString(1, userName);
rs = pst.executeQuery();
if (rs.next()) {
int userId = rs.getInt("userId");
String userPwd = rs.getString("UserPwd");
String userType = rs.getString("userType");
User user = new User();
user.setUserId(userId);
user.setUserName(userName);
user.setUserPwd(userPwd);
```

```
user. setUserType(userType);
return user;
}
} catch (SQLException e) {
System. out. println("数据库操作失败!");
} finally {
DBUtil. release(conn, pst, rs);
}
return null;
}
public boolean insertUser(String userName, String userPwd, String userType) {
String sql = " insert into edu _ user ( userName, userPwd, userType ) values(?,?,?)";
try {
conn = DBUtil. getConnection( );
pst = conn. prepareStatement(sql);
pst. setString(1, userName);
pst. setString(2, userPwd);
pst. setString(3, userType);
pst. executeUpdate( );
return true;
} catch (SQLException e) {
System. out. println("数据库操作失败!");
System. out. println(e. toString( ));
e. printStackTrace( );
} finally {
DBUtil. release(conn, pst, null);
}
return false;
}
}
```

StudentDaoImpl 类的部分代码如下所示。

```
//StudentDaoImpl. java
……
public class StudentDaoImpl implements StudentDao {
```

```
Connection conn = null;
PreparedStatement pst = null;
Statement st = null;
ResultSet rs = null;
public List<Student> queryAllStu( ) {
String sql = "select * from edu_stu";
List<Student> list = new ArrayList<Student>( );
try {
conn = DBUtil. getConnection( );
st = conn. createStatement( );
rs = st. executeQuery(sql);
while (rs. next( )) {
Student stu = new Student( );
stu. setStuNum(rs. getString("stuNum"));
stu. setStuName(rs. getString("stuName"));
stu. setStuSex(rs. getString("stuSex"));
stu. setStuAge(rs. getInt("stuAge"));
stu. setStuMajor(rs. getString("stuMajor"));
list. add(stu);
}
return list;
} catch (SQLException e) {
System. out. println("数据库操作失败!");
} finally {
DBUtil. release(conn, st, rs);
}
return null;
}
public List<Student> queryAllStuByCondition(String condition) {
String sql = "select * from edu_stu where " + condition;
List<Student> list = new ArrayList<Student>( );
try {
conn = DBUtil. getConnection( );
st = conn. createStatement( );
rs = st. executeQuery(sql);
```

```
while (rs.next( )) {
Student stu = new Student( );
stu.setStuNum(rs.getString("stuNum"));
stu.setStuName(rs.getString("stuName"));
stu.setStuSex(rs.getString("stuSex"));
stu.setStuAge(rs.getInt("stuAge"));
stu.setStuMajor(rs.getString("stuMajor"));
list.add(stu);
}
return list;
} catch (SQLException e) {
System.out.println("数据库操作失败!");
} finally {
DBUtil.release(conn, st, rs);
}
return null;
}
public Student queryStuByStuNum(String stuNum) {
String sql = "select * from edu_stu where stuNum=?";
try {
conn = DBUtil.getConnection( );
pst = conn.prepareStatement(sql);
pst.setString(1, stuNum);
rs = pst.executeQuery( );
Student stu = new Student( );
if (rs.next( )) {
stu.setStuNum(rs.getString("stuNum"));
stu.setStuName(rs.getString("stuName"));
stu.setStuSex(rs.getString("stuSex"));
stu.setStuAge(rs.getInt("stuAge"));
stu.setStuMajor(rs.getString("stuMajor"));
return stu;
}
} catch (SQLException e) {
System.out.println("数据库操作失败!");
```

```
} finally {
DBUtil. release(conn, st, rs);
}
return null;
}
}
```

(6)定义业务逻辑实现类service。在工程中新建cn. hbmy. p10_1. service包,在包中新建service业务逻辑接口UserService和StudentService。User类的业务操作包括checkUser(登录验证)和registUser(用户注册)。Student类的业务操作包括queryAllStuS(查询所有学生记录)、queryStuByStuNumS(按学号查询学生记录)和queryAllStuByConditionS(按条件查询学生记录)。UserService接口代码如下所示。

```
//UserService. java
……
public interface UserService {
public boolean checkUser(User user);
public boolean registUser(User user);
}
```

StudentService接口代码如下所示。

```
//StudentService. java
……
public interface StudentService {
public List<Student> queryAllStuS( );
public List<Student> queryAllStuByConditionS(String condition);
public Student queryStuByStuNumS(String stuNum);
}
```

(7)建立service接口实现类。在工程中新建cn. hbmy. p10_1. service. impl包,在包中新建service业务逻辑接口的实现类UserServiceImpl和StudenServiceImpl。UserServiceImpl类代码如下所示。

```
//UserServiceImpl. java
……
public class UserServiceImpl implements UserService {
UserDao userDao=new UserDaoImpl( );
public boolean checkUser(User user) {
UserDao userDao=new UserDaoImpl( );
User u=userDao. queryByUserName(user. getUserName( ));
```

```
if(u! =null) {
String p1=user. getUserPwd( );
String p2=u. getUserPwd( );
String t1=user. getUserType( );
String t2=u. getUserType( );
if(p1. equals(p2) && t1. equals(t2))
return true;
else
return false;
}else
return false;
}
public boolean registUser(User user) {
UserDao userDao=new UserDaoImpl( );
User u=userDao. queryByUserName(user. getUserName( ));
if(u==null){
userDao. insertUser ( user. getUserName ( ), user. getUserPwd ( ), user. getUserType( ));
return true;
}else
return false;
}
}
```

StudenServiceImpl 类代码如下所示。

```
//UserServiceImpl. java
……
public class StudentServiceImpl implements StudentService {
public List<Student> queryAllStuS( ) {
StudentDao studentDao=new StudentDaoImpl( );
List<Student> list=new ArrayList<Student>( );
list=studentDao. queryAllStu( );
return list;
}
public List<Student> queryAllStuByConditionS(String condition) {
StudentDao studentDao=new StudentDaoImpl( );
```

```
List<Student> list=new ArrayList<Student>( );
list=studentDao. queryAllStuByCondition(condition);
return list;
}
public Student queryStuByStuNumS(String stuNum) {
StudentDao studentDao=new StudentDaoImpl( );
Student stu=studentDao. queryStuByStuNum(stuNum);
return stu;
}
}
```

注意体会 Service 和 Dao 接口的不同，以及 ServiceImpl 与 DaoImpl 的差异。

(8)建立 Action 响应用户请求。在工程中新建 cn. hbmy. p10_1. action 包，在包中新建响应用户登录验证请求的 LoginAction，响应用户注册请求的 RegistAction，响应查询请求的 QueryStuAction。具体是查询所有学生，还是按条件查询，或是按照学号查询都在 QueryStuAction 中处理。LoginAction 的代码如下所示。

```
//LoginAction. java
……
public class LoginAction extends ActionSupport {
private String userName;
private String userPwd;
private String userType;
//setter 和 getter 方法省略
……
public HttpServletRequest request;
public String execute( ) throws Exception {
UserService userService=new UserServiceImpl( );
User user=new User( );
if(!"". equals(userName) && !"". equals(userPwd) && "". equals(userPwd)){
user. setUserName(userName);
user. setUserPwd(userPwd);
user. setUserType(userType);
if(userService. checkUser(user)){
request=ServletActionContext. getRequest( );
request. getSession( ). setAttribute("user-info",user );
return "SUCCESS";
```

```
}
}
return "ERROR";
}
}
```

注意,这里通过 setAttribute("user—info",user)方法将用户信息存入 session 范围内,可用于用户访问权限验证,本工程并未完成该功能,读者可尝试完成该功能。

RegistAction 的代码如下所示。

```
//RegistAction.java
……
public class RegistAction extends ActionSupport {
private String userName;
private String userPwd;
//省略 setter 和 getter 方法
……
public HttpServletRequest request;
public String execute( ) throws Exception {
UserService userService=new UserServiceImpl( );
User user=new User( );
boolean b=false;
if (!"". equals ( userName ) && !"". equals ( userPwd ) && !"". equals (userType)){
user.setUserName(userName);
user.setUserPwd(userPwd);
user.setUserType(userType);
b= userService.registUser(user);
if (b)
return "LOGIN";
}
return "ERROR";
}
……
```

QueryStuAction 的代码如下所示。

```
//QueryStuAction.java
……
```

```
public HttpServletRequest request;
public String execute( ) throws Exception {
request=ServletActionContext.getRequest( );
StudentService stuService=new StudentServiceImpl( );
String param=request.getParameter("param");
String value1=request.getParameter("value1");
String op=request.getParameter("op");
String field=request.getParameter("field");
String stuNum=request.getParameter("stuNum");
String condition="";
if("stuAge".equals(field))
condition=field+op+value1;
else
condition=field+op+"'"+value1+"'";
List<Student> list=new ArrayList<Student>( );
if("all".equals(param)){//查询所有记录
list=stuService.queryAllStuS( );
request.getSession( ).setAttribute("dataList", list);
}
if("byCdn".equals(param) && !"".equals(field) && !"".equals(op) &&
!"".equals(value1) ){//按条件查询记录
list=stuService.queryAllStuByConditionS(condition);
request.getSession( ).setAttribute("dataList", list);
}
if("byNum".equals(param)){//按学号查询记录
Student stu=stuService.queryStuByStuNumS(stuNum);
list.add(stu);
request.getSession( ).setAttribute("dataList", list);
}
return "SUCCESS";
}
……
```

注意，在 QueryStuAction 中只是根据请求，调用相应的 Service 的业务逻辑，再由 Service 业务逻辑调用 DAO 数据访问逻辑，最后再通过 return 实现跳转。

(9)建立 JSP 用户请求页面。在工程的 WebContent 目录下新建一个 JSP 页面。

login. jsp 用于用户登录和注册;error. jsp 用于错误请求响应的页面;queryStu. jsp 用于显示查询结果。login. jsp 页面部分代码如下。

```
<!--login.jsp-->
……
<%@taglib prefix="s" uri="/struts-tags"%>
……
<s:i18n name="globalMessages">
<center>
<s:text name="check"></s:text>
<a href="switch.action? request_locale=zh_CN">
<s:text name="chinese"></s:text></a>
<a href="switch.action? request_locale=en_US"><s:text name="english">
</s:text></a>
<h3><s:text name="title"></s:text></h3>
<s:form action="login.action" method="post" name="myform">
<table>
<tr><td><s:textfield name="userName" key="uname" /></td></tr>
<tr><td><s:password name="userPwd" key="upwd" /></td></tr>
<tr> <td><input type="submit" value="登录" id="login"
onclick="myform.action='login.action';myForm.submit( )"/></td>
<td><input type="submit" value="注册" id="regist"
onclick="myform.action='regist.action';myForm.submit( )"/></td></tr>
……
```

注意 login. jsp 页面中使用了国际化操作,后面将进一步说明。error. jsp 页面部分代码如下。

```
<!-- error.jsp-->
……
<body>
登录或注册失败! 请点击<a href="login.jsp">这里</a>重新登录! 或者 5 秒后将
自动跳转到登录页面! <%response.setHeader("refresh","5;URL=login.jsp"); %>
</body>
……
```

queryStu. jsp 页面部分代码如下。

```
<!-- queryStu.jsp-->
<%@ page contentType="text/html; charset=UTF-8" pageEncoding="UTF
```

```
-8"%>
    <%@ taglib uri="http://java.sun.com/jsp/jstl/core" prefix="c"%>
    <%@taglib prefix="s" uri="/struts-tags"%>
    ......
    <link rel="stylesheet" type="text/css" href="css/tb.css">
    ......
    <s:form action="queryStu.action?param=all" method="post">
    点击查看所有学生记录:<input type="submit" value="查看记录" />
    </s:form>
    <s:form action="queryStu.action?param=byNum" method="post">
    输入学号查询:<input type="text" name="stuNum" />
    <input type="submit" value="查看记录" />
    </s:form>
    <s:form action="queryStu.action?param=byCdn" method="post">
    输入条件查询:
    <s:select list="#{'stuNum':'学号','stuName':'姓名','stuSex':'性别','
stuAge':'年龄','stuMajor':'专业'}"
    headerKey="" headerValue="请选择字段" theme="simple" name="field">
</s:select>
    <s:select list="#{'>':'>','<':'<','=':'='}" theme="simple" name="
op"/>
    <s:textfield theme="simple" name="value1" />
    <input type="submit" value="查看记录" />
    </s:form>
    <table id="students">
    <tr>
    <th>学号</th>
    <th>姓名</th>
    <th>性别</th>
    <th>年龄</th>
    <th>专业</th>
    </tr>
    <c:forEach var="stu" items="${dataList}">
    <tr class='alt'>
    <td>${stu.stuNum}</td>
```

```
<td>${stu.stuName}</td>
<td>${stu.stuSex}</td>
<td>${stu.stuAge}</td>
<td>${stu.stuMajor}</td>
……
```

(10)完成 Struts 2 的 xml 配置。在 struts.xml 文件中完成对 action 的配置，struts.xml 中注册了工程需要的所有 action 类。Eclipse 不支持直接生成 Struts2 的配置文件，可手动设置添加该配置文件。从下载的 Struts 2 的开发文件夹中找到 Strut 2 的 core 包，如 struts2－core－2.5.16，解压缩后找到 dtd 文件，如 struts－2.5.dtd。然后在 Eclipse 中进行配置。点击 Windows 菜单，选择 Peference | XML | XML Catalog，打开如图 10－4 所示的对话框，单击 Add，弹出如图 10－5 所示对话框并进行相关设置。

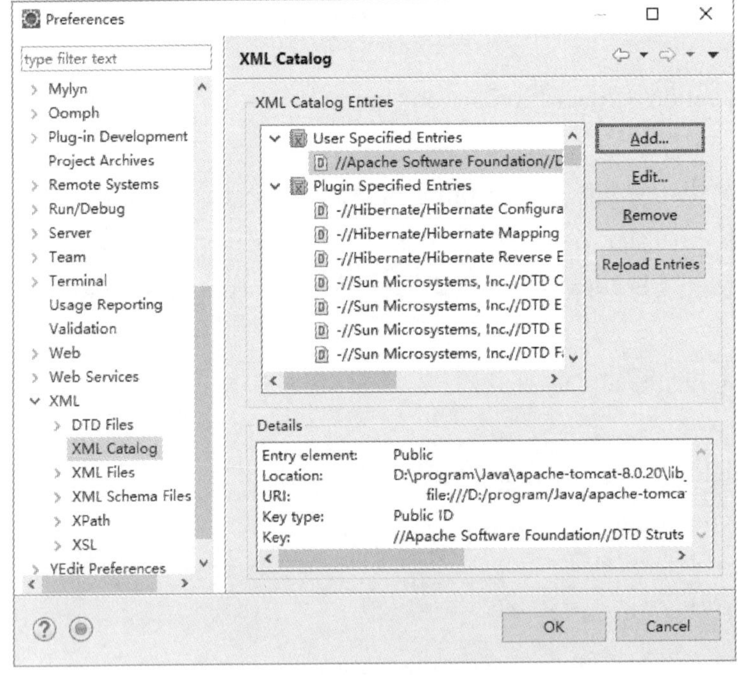

图 10－4 XML Catalog 设置

在 Location 处选择 struts－2.5.dtd 文件的路径，在 Key type 处选择 public ID，在 Key 处设置为：－//Apache Software Foundation//DTD Struts Configuration 2.5//EN，在 Alternative web address 处设置为：http://struts.apache.org/dtds/struts－2.5.dtd。后面 2 个信息的具体内容，可以在 struts－2.5.dtd 文件中找到。完成设置后，就可以在 Eclipse 中新建 Struts.xml 文件了。

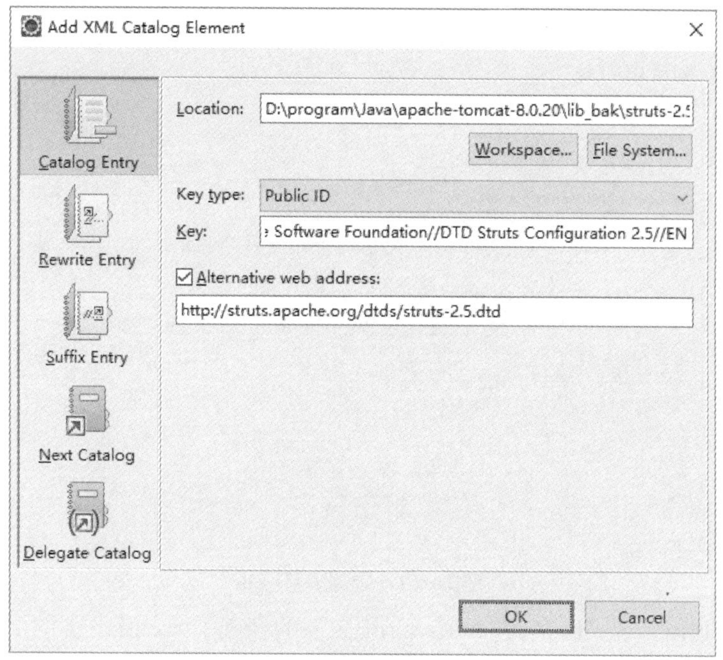

图 10—5 添加指定的 dtd 文件

(11)在 src 目录下新建 Struts. xml 文件。右击 src 目录,在弹出快捷菜单中选择 New | Other ,弹出对话框中选择 XML | XML File,单击 Next 按钮,弹出对话框做如图 10—6 所示的选择。

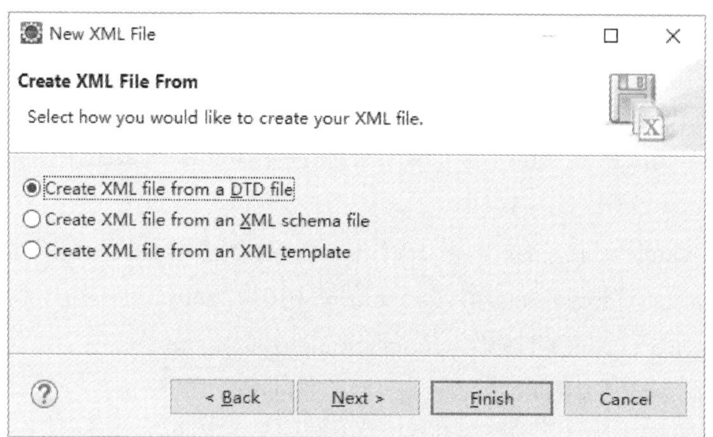

图 10—6 选择生成 XML 文件方式

单击 Next 按钮,弹出对话框中做如图 10—7 的字的选择。

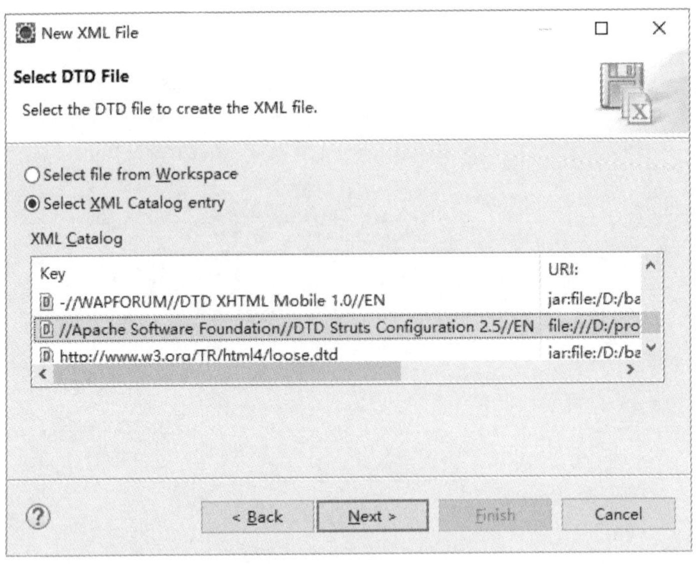

图 10—7　选择 DTD File

最后单击 Finish 按钮，完成 struts. xml 文件的新建。src 目录下的 struts. xml 文件在编译后会自动生成相应文件至工程的 class 文件夹中。

(12)配置 struts. xml 文件。具体代码如下所示。

```
<!--struts.xm-->
<?xml version="1.0" encoding="UTF-8"?>
<!DOCTYPE struts PUBLIC "//Apache Software Foundation//DTD Struts Configuration 2.5//EN" "http://struts.apache.org/dtds/struts-2.5.dtd">
<struts>
<package name="default" namespace="/" extends="struts-default">
</package>
<package name="loginandregister" namespace="/" extends="default">
<action name="login" class="cn.hbmy.p10_1.action.LoginAction">
<result name="SUCCESS">/queryStu.jsp</result>
<result name="ERROR">/error.jsp</result>
</action>
<action name="regist" class="cn.hbmy.p10_1.action.RegistAction">
<result name="LOGIN">/login.jsp</result>
<result name="ERROR">/error.jsp</result>
</action>
</package>
<package name="userrights" namespace="/" extends="default">
<action name="queryStu" class="cn.hbmy.p10_1.action.QueryStuAction">
```

```
<result name="SUCCESS">/queryStu. jsp</result>
</action>
</package>
</struts>
```

注意，可以在 struts. xml 文件中对 action 进行注册，也可以通过注解的方式直接在定义 Action 时进行注册。读者也可尝试通过注解方式注册 action。

(13)配置 web. xml 文件。在 WebContent | WEB INF 目录下建立 web. xml 文件，用于装载 Struts 2 过滤器，并设置欢迎页面为 login. jsp。具体代码如下所示。

```
<! --struts. xml-->
<? xml version="1. 0" encoding="UTF-8"? >
<web-app xmlns:xsi="http://www. w3. org/2001/XMLSchema-instance"
xmlns="http://java. sun. com/xml/ns/javaee"
xsi: schemaLocation = " http://java. sun. com/xml/ns/javaee http://java. sun.
com/xml/ns/javaee/web-app_3_0. xsd"
id="WebApp_ID" version="3. 0">
<display-name>Struts2 demo</display-name>
<filter>
<filter-name>struts2</filter-name>
<filter-class>
org. apache. struts2. dispatcher. filter. StrutsPrepareAndExecuteFilter
</filter-class>
</filter>
<filter-mapping>
<filter-name>struts2</filter-name>
<url-pattern>/ * </url-pattern>
</filter-mapping>
<welcome-file-list>
<welcome-file>/login. jsp</welcome-file>
</welcome-file-list>
</web-app>
```

(14)为工程创建国际化资源文件。这里以登录页面为例实现国际化，新建 resource 目录，并在目录下新建 2 个国际化资源文件。创建 messages_zh_CN. properties，文件代码如下所示。

```
uname=用户名
upwd=密码
```

```
utype=用户类型
login=登录
regist=注册
title=欢迎
teacher=老师
student=学生
check=请选择语言
english=英语
chinese=中文
```

创建 messages_en_US. properties，文件代码如下所示。

```
uname=user name
upwd=password
utype=user type
login=login
regist=register
title=welcome
teacher=teacher
student=student
check=please check your language
english=English
chinese=Chinese
```

注意，资源文件的命名必须符合命名规则。以 messages_zh_CN. properties 为例，messages 为资源文件的基本名，可自行定义，也对应了 struts. xml 文件中的定义；zh_CN 分别表示语言和国家，这个必须是 Java 所支持的语言和国家，且不能随意修改。

(15)为了在登录页面中实现国际化，当选择 English | Chinese 时将显示不同页面语言，这里需要新增 SwitchAction，用于给国际化拦截器传递 request_locale 参数，从而确定采用哪种语言。在 cn. hbmy. p10_1. action 包中，新建 SwitchAction 类，代码如下所示。

```
//SwitchAction. java
package cn. hbmy. p10_1. action;
import com. opensymphony. xwork2. ActionSupport;
public class SwitchAction extends ActionSupport {
private static final long serialVersionUID = 1L;
public String execute( ) throws Exception {
return "LOGIN";
```

```
}
}
```

请思考通过该 Action 是如何实现页面的国际化的？另外，在 Struts. xml 文件中添加相关 Action 的部署和国际化拦截器，如下所示。

```
<constant name="struts. custom. i18n. resources" value="messages"/>
……
<action name="switch" class="cn. hbmy. p10_1. action. SwitchAction">
<result name="LOGIN">/login. jsp</result>
</action>
```

其中 value="messages"用于加载国际化资源文件。

(16)程序运行效果。登录页面如图 10—8 所示。

图 10—8 用户登录页面

当输入错误的登录信息或者错误的注册信息时(错误注册信息包括注册用户名、用户类型或密码为空，或者用户名已经存在等)，这些情况都将返回 error. jsp 错误处理页面，如图 10—9 所示。

图 10—9 错误页面

注册时，如果输入的注册信息正确，则正常保存注册信息，并要求登录。当输入正确登录信息时，将进入查询页面，进行数据查询。如图 10—10 所示。

学号	姓名	性别	年龄	专业
031640101	张丽	女	20	计算机科学与技术
031640102	李立	男	20	计算机科学与技术
031640103	杨文	男	20	计算机科学与技术
031640201	王琴	女	19	数字媒体
031640202	徐方	男	21	数字媒体
031640203	覃钟	男	20	数字媒体
031640301	李强	男	19	信息安全
031640302	朱雨	女	20	信息安全

图 10—10　查询数据页面

四、实验思考

1. 按步骤完成实验内容，对每个步骤截图并保存，形成实验报告。

2. 在本实验基础上，给程序添加用户身份验证的 filter 和解决 post 提交方式乱码的 filter。

3. 思考角色的权限控制问题，尝试提出角色的权限控制的解决方案。

4. 思考如何对登录时输入的用户信息进行有效性检验。例如，输入用户名和密码必须符合长度要求，必须是合法字符等，提出解决方案并尝试实现。

实验十一　Spring MVC 开发框架

一、实验目的

1. 掌握 Spring MVC 框架概念及工作原理。
2. 掌握 Spring MVC 访问数据库连接池的方法。
3. 理解和掌握 Spring MVC 通过注解实现 Bean 的自动装配方法。
4. 理解和掌握 Spring MVC 视图解析器的作用和配置方法。

二、基础知识

1. Spring MVC 框架。

Spring MVC 提供了一种 MVC 模式的结构和组件,它是属于 Spring 框架的后续产品,是 Spring 框架提供的构建 Web 应用程序的全功能 MVC 模块。在使用 Spring 进行 WEB 开发时,可以选择使用 Spring 的 Spring MVC 框架或集成其他 MVC 开发框架,比如 Struts2。

2. Spring MVC 工作原理。

(1)客户端发送请求至前端控制器 DispatcherServlet。

(2)DispatcherServlet 收到请求,调用 HandlerMapping 处理器映射器。

(3)处理器映射器根据请求 url 找到具体的处理器,生成处理器对象及处理器拦截器(如果有则生成)一并返回给 DispatcherServlet。

(4)DispatcherServlet 通过 HandlerAdapter 处理器适配器调用处理器。

(5)执行处理器(Controller,或叫后端控制器)。

(6)Controller 执行完成后返回 ModelAndView。

(7) HandlerAdapter 将 controller 执行结果 ModelAndView 返回给 DispatcherServlet。

(8)DispatcherServlet 将 ModelAndView 传给 ViewReslover 视图解析器。

(9)ViewReslover 解析后返回具体 View。

(10)DispatcherServlet 对 View 进行渲染视图(即将模型数据填充至视图中)。

(11)DispatcherServlet 响应用户。

3. Spring MVC 常用注解。

(1)@Controller。@Controller 用于标记在一个类上,使用它标记一个 Spring MVC 的 Controller 对象。控制器 Controller 负责处理由 DispatcherServlet 分发的请求,它把用户请求的数据经过业务处理层处理之后封装成一个 Model ,然后再把该 Model 返回给对应的 View 进行展示。

(2)@RequestMapping。@RequestMapping 是一个用来处理请求地址映射的注解,可用于类或方法。用于类,表示类中的所有响应请求的方法都是以该地址作为父路径;用于方法,表示该方法用于响应请求。

(3)@Autowired。@Autowired 用于 bean 的注入,用于属性或 setter 方法,作用类似@Resource。

(4)@Component。@Component 用于不知道某类属于哪一层时,但不建议这样使用。

(5)@Repository。@Repository 用于注解 dao 层的类,通常用于注解 dao 的实现类。

(6)@Service。@Service 用于注解 service 层的类,通常用于注解 service 的实现类。

4. Bean 的自动载入。

通过在 Spring MVC 的配置文件中设置 context:component—scan 可自动扫描指定

区域 base－package 里注解的 Java 类，并将它们自动注册为 Bean，支持的注解方式包括@Component、@Controller、@Service 和@Reposity。

三、实验步骤

1. 下载 Spring 开发 jar 包。

(1)在 http://maven.springframework.org/release/org/springframework/spring/网站下载最新 Spring 框架开发 jar 包，如本书实验所下载的 spring－framework－5.0.5.RELEASE－dist.zip。

(2)将下载的压缩包解压，在得到的文件中，libs 目录下就是 Spring5 框架依赖的 jar 包文件。其中本实验依赖的 Spring MVC 的核心 jar 包见表 11－1。

表 11－1　Spring5 的核心 jar 包文件

序号	包名
1	spring－beans－5.0.5.RELEASE.jar
2	spring－context－5.0.5.RELEASE.jar
3	spring－core－5.0.5.RELEASE.jar
4	spring－web－5.0.5.RELEASE.jar
5	spring－webmvc－5.0.5.RELEASE.jar
6	spring－aop－5.0.5.RELEASE.jar
7	spring－expression－5.0.5.RELEASE.jar
8	spring－tx－5.0.5.RELEASE.jar
9	spring－aop－5.0.5.RELEASE.jar
10	spring－jdbc－5.0.5.RELEASE.jar

2. 新建一个 Web 工程，使用 Spring MVC 框架实现工程 p10_1 的 Web 数据库登录访问。由于 Struts 和 Spring MVC 都是 MVC 模式框架，只是 Spring MVC 在控制层的流程控制逻辑更加灵活，因此除了控制层外，本工程中的部分文件代码可参照工程 p10_1。

(1)在 Eclipse 中，新建一个 Web 工程 p11_1。在工程中导入 Spring MVC 的核心开发包和 MySQL 数据库驱动包，数据库连接池 Druid，以及其他相关 jar 包，导入 jar 包到 p11_1/WebContent/WEB－INF/lib 目录下，如图 11－1 所示。

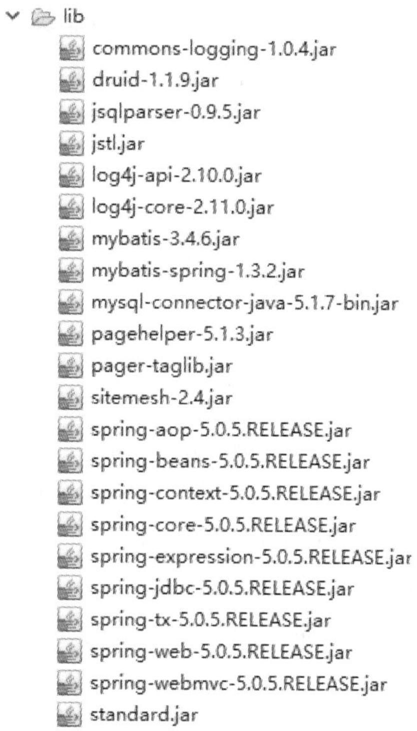

```
lib
  commons-logging-1.0.4.jar
  druid-1.1.9.jar
  jsqlparser-0.9.5.jar
  jstl.jar
  log4j-api-2.10.0.jar
  log4j-core-2.11.0.jar
  mybatis-3.4.6.jar
  mybatis-spring-1.3.2.jar
  mysql-connector-java-5.1.7-bin.jar
  pagehelper-5.1.3.jar
  pager-taglib.jar
  sitemesh-2.4.jar
  spring-aop-5.0.5.RELEASE.jar
  spring-beans-5.0.5.RELEASE.jar
  spring-context-5.0.5.RELEASE.jar
  spring-core-5.0.5.RELEASE.jar
  spring-expression-5.0.5.RELEASE.jar
  spring-jdbc-5.0.5.RELEASE.jar
  spring-tx-5.0.5.RELEASE.jar
  spring-web-5.0.5.RELEASE.jar
  spring-webmvc-5.0.5.RELEASE.jar
  standard.jar
```

图 11—1　工程导入 jar 包

(2)参照实验十的工程 p10_1 完成以下操作。在工程 p11_1 中新建 cn. hbmy. p11_1. entity 包,在包中新建 Student 和 User 类;新建 cn. hbmy. p11_1. dao 包,在包中新建 IStudentDao 接口和 IUserDao 接口,以及 StudentDao 和 UserDao 接口实现类,其中 2 个接口对应 p10_1 中的 StudentDao 和 UserDao 接口,2 个接口实现类对应 p10_1 中的 StudentDaoImpl 和 UserDaoImpl 类;新建 cn. hbmy. p11_1. Service 包,在包中新建 IStudentService 和 IUserService 接口,以及 StudentService 和 UserService 接口实现类,其中 2 个接口对应于 p10_1 中的 StudentService 和 UserService 接口,2 个接口实现类对应 p10_1 中的 StudentServiceImpl 和 UserServiceImpl 类。完成后的包文件结构如图 11—2 所示。

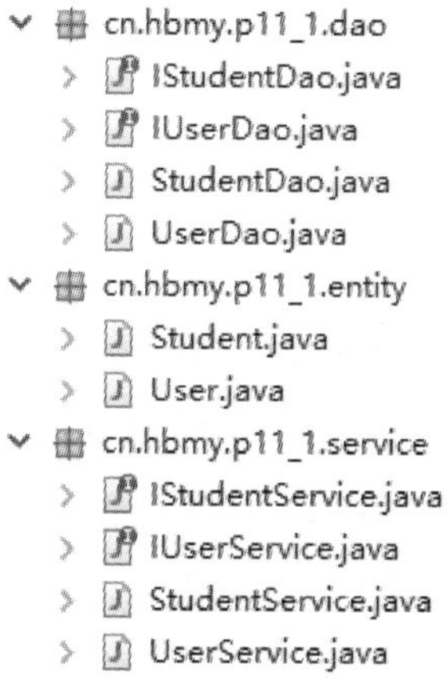

```
cn.hbmy.p11_1.dao
  IStudentDao.java
  IUserDao.java
  StudentDao.java
  UserDao.java
cn.hbmy.p11_1.entity
  Student.java
  User.java
cn.hbmy.p11_1.service
  IStudentService.java
  IUserService.java
  StudentService.java
  UserService.java
```

图 11—2　工程部分包文件结构

(3)重点改造 Spring MVC 的控制层。在工程中新建 cn. hbmy. p11_1. controller 包,在包中新建 UserController 类,用于响应用户的登录验证请求和注册请求,类代码如下所示。

```
//UserController. java
……
@Controller
public class UserController {
@Autowired
private UserService userService;
@RequestMapping(value="/login. do",method=RequestMethod. POST)
public String login(@RequestParam("username")String username,
@RequestParam("userpwd")String userpwd,
@RequestParam("usertype")String usertype,
Model model) throws Exception {
User user=new User( );
boolean b=false;
if(!"". equals(username) && !"". equals(userpwd) && !"". equals(usertype))
{
user. setUserName(username);
user. setUserPwd(userpwd);
user. setUserType(usertype);
b= userService. checkUser(user);
}
if (b) {
model. addAttribute(user);
return "student/query";
}
return "error";
}
@RequestMapping(value="/regist. do",method=RequestMethod. POST)
public String regist(@RequestParam("username")String username,
@RequestParam("userpwd")String userpwd,
@RequestParam("usertype")String usertype) throws Exception {
User user=new User( );
boolean b=false;
```

```
if(!"".equals(username) && !"".equals(userpwd) && !"".equals(usertype))
{
user.setUserName(username);
user.setUserPwd(userpwd);
user.setUserType(usertype);
b= userService.registUser(user);
if (b)
return "login";
}else
return "error";
return "error";
}
@RequestMapping(value="/newlogin.do",method=RequestMethod.GET)
public String callLogin( ) throws Exception {
return "login";
}
}
```

在 UserController 类中，通过注解@Controller，标注了类的类型是 Controller；注解@RequestMapping(value= "/login. do", method= RequestMethod. POST)，用于请求地址映射，可用于类或方法上，用于类上即是对类中所有的响应请求的方法都以该映射路径为父路径。这里是用于方法上，表示请求的映射路径是“/login. do”，并指定了请求类型为 request；注解@RequestParam（"username"），用于表明传递的请求参数为 username；注解@Autowired，用于装配 bean，注入属性，功能类似 JDK 支持的注解@Resource。

当处理完请求，通过 return 转发到下一资源，可以是下一个 Controller 或者 View，这里是转发到 WEB－INF 下的 jsp 文件，与 dispatch－servlet. xml 配置视图解析器一同作用。

(4) 完成另一个 Controller 的设计。在 cn. hbmy. p11_1. controller 包中新建 StudentController 类，用于响应所有针对 Student 的请求，类代码如下所示。

```
//StudentController.java
……
@Controller
public class StudentController {
@Autowired
private StudentService studentService;
```

```
@RequestMapping(value="/queryAll.do",method=RequestMethod.POST)
public String queryAll(Model model) throws Exception {
List<Student> list=studentService.queryAllStuS();
model.addAttribute("stus", list);
return "student/query";
}
@RequestMapping(value="/queryByStuNum.do", method=RequestMethod.POST)
public String queryByStuNum(@RequestParam("stuNum")String stuNum,Model model) throws Exception {
Student stu=studentService.queryStuByStuNumS(stuNum);
List<Student> stus=new ArrayList<Student>();
stus.add(stu);
model.addAttribute("stus", stus);
return "student/query";
}
@RequestMapping(value="/queryByCondition.do", method=RequestMethod.POST)
public String queryByCondition(@RequestParam("field") String field, @RequestParam("op") String op, @RequestParam("value")String value,Model model) throws Exception {
String condition="";
if("stuAge".equals(field))
condition=field+op+value;
else
condition=field+op+"'"+value+"'";
List<Student> list=studentService.queryAllStuByConditionS(condition);
model.addAttribute("stus", list);
return "student/query";
}
}
```

(5)在工程中新建 Spring 的配置文件。右击 src 目录，在弹出的快捷菜单中选择 New | Other，在弹出对话框中选择 Spring | Spring Bean Configuration File，点击 Next 按钮，填写配置文件名为 applicationContext.xml，点击 Finish 按钮完成配置文件的创建，该配置文件代码如下所示。

```
<! --applicationContext. xml-->
<? xml version="1. 0" encoding="UTF-8"? >
<beans xmlns="http://www. springframework. org/schema/beans"
xmlns:xsi="http://www. w3. org/2001/XMLSchema-instance"
xmlns:context="http://www. springframework. org/schema/context"
xmlns:batch=" http://www. springframework. org/schema/batch"
xsi:schemaLocation="http://www. springframework. org/schema/context
http://www. springframework. org/schema/context/spring-context-4. 2. xsd
http://www. springframework. org/schema/beans http://www. springframework. org/
schema/beans/spring-beans-4. 2. xsd">
<bean id="property"
class="org. springframework. beans. factory. config. PropertyPlaceholderConfigurer">
<property name="locations">
<list>
<! -- classpath 即 src 根目录 -->
<value>classpath:jdbc. properties</value>
</list>
</property>
</bean>
<! --启动自动扫描 -->
<context:component-scan base-package="cn. hbmy. p11_1">
<! -- 不扫描注解为 controller 的类型 -->
<context:exclude-filter type="annotation"
expression="org. springframework. stereotype. Controller" />
</context:component-scan>
<! --默认的 spring 容器的注解映射的支持 -->
<context:annotation-config />
<bean id="dataSource"
class="com. alibaba. druid. pool. DruidDataSource" destroy-method="close">
<! -- 指定连接数据库的 JDBC 驱动 -->
<property name="driverClassName" value="${driver}" />
<! -- 连接数据库所用的 URL -->
<property name="url" value="${url}" />
<! -- 连接数据库的用户名 -->
<property name="username" value="${username}" />
```

```
<! —— 连接数据库的密码 ——>
<property name="password" value="${password}" />
<! —— 配置初始化大小、最小、最大 ——>
<property name="initialSize" value="1" />
<property name="minIdle" value="1" />
<property name="maxActive" value="10" />
<! —— 配置获取连接等待超时的时间 ——>
<property name="maxWait" value="10000" />
<! —— 配置间隔多久进行一次检测,检测需要关闭的空闲连接,单位是毫秒 —
—>
<property name="timeBetweenEvictionRunsMillis" value="60000" />
<! —— 配置一个连接在池中最小生存的时间,单位是毫秒 ——>
<property name="minEvictableIdleTimeMillis" value="25200000" />
<! —— 申请连接的时候检测 ——>
<property name="testWhileIdle" value="true" />
<! —— 配置强制关闭长时间不使用的连接 ——>
<property name="removeAbandoned" value="true" />
<! —— 配置超时断开空闲连接的时间, 单位是秒——>
<property name="removeAbandonedTimeout" value="1800" />
<! —— 关闭 abanded 连接时输出错误日志 ——>
<property name="logAbandoned" value="true" />
<! —— 监控数据库 ——>
<property name="filters" value="stat" />
</bean>
</beans>
```

(6)在工程中新建 Spring MVC 的配置文件。通常将 applicationContext. xml 文件作为 Spring 容器的配置文件,而对 Spring MVC 的配置文件则另外新建。在工程中的 WEB—INF 文件夹下新建 dispatch—servlet. xml 文件,该配置文件代码如下所示。

```
<! ——dispatch—servlet. xml——>
<? xml version="1. 0" encoding="UTF—8"? >
<beans xmlns="http://www. springframework. org/schema/beans"
xmlns:xsi="http://www. w3. org/2001/XMLSchema—instance"
xmlns:context="http://www. springframework. org/schema/context"
xmlns:mvc="http://www. springframework. org/schema/mvc"
xsi:schemaLocation="http://www. springframework. org/schema/beans http://
```

```
www.springframework.org/schema/beans/spring－beans.xsd
    http://www.springframework.org/schema/context http://www.springframework.org/schema/context/spring－context－4.3.xsd
    http://www.springframework.org/schema/mvchttp://www.springframework.org/schema/mvc/spring－mvc－4.3.xsd">
    <!－－用于 springmvc 的注解支持 －－>
    <mvc:annotation－driven />
    <!－设置访问静态资源 －－>
    <mvc:resources location="/css/" mapping="/css_mapping/**"/>
    <!－－启动自动扫描 －－>
    <context:component－scan base－package="cn.hbmy.p11_1">
    <!－－只扫描@Controller 注解的 JAVA 类 －－>
    <context:include－filter type="annotation"
    expression="org.springframework.stereotype.Controller" />
    </context:component－scan>
    <!－－配置视图解析器 －－>
    <bean class="org.springframework.web.servlet.view.UrlBasedViewResolver">
    <property name="viewClass" value="org.springframework.web.servlet.view.JstlView"/>
    <property name="prefix" value="/WEB－INF/"/>
    <property name="suffix" value=".jsp"/>
    </bean>
    </beans>
```

注意,dispatch－servlet. xml 文件作为 Spring MVC 的配置文件,需要使用 spring－mvc. xsd。

(7)装载工程配置文件 web. xml。整个工程以 Spring MVC 作为 Web 框架,通过 Spring 访问 Druid 数据库连接池进行数据库连接,因此需要在 web. xml 文件中分别加载 Spring 配置文件和 Spring MVC 配置文件,同时装载 DruidWebStatFilter 实现连接池监控,装载 encodingFilter 实现请求字符的统一编码。完整的 web. xml 文件代码如下所示。

```
<!－web.xml－－>
<?xml version="1.0" encoding="UTF－8"?>
<web－app
xmlns:xsi="http://www.w3.org/2001/XMLSchema－instance" xmlns="http://xmlns.jcp.org/xml/ns/javaee" xsi:schemaLocation="http://xmlns.jcp.org/
```

```
xml/ns/javaee http://xmlns.jcp.org/xml/ns/javaee/web-app_3_1.xsd" id="
WebApp_ID" version="3.1">
    <display-name>p11_1</display-name>
      <!--加载配置文件-->
        <context-param>
            <param-name>contextConfigLocation</param-name>
              <param-value>classpath*:applicationContext.xml</param-
value>
        </context-param>
        <listener>
    <listener-class>
    org.springframework.web.util.IntrospectorCleanupListener
    </listener-class>
    </listener>
        <listener>
    <listener-class>
    org.springframework.web.context.ContextLoaderListener
    </listener-class>
        </listener>
        <!-- spring mvc -->
        <servlet>
            <servlet-name>springMvc</servlet-name>
            <servlet-class>
    org.springframework.web.servlet.DispatcherServlet
    </servlet-class>
            <init-param>
                <param-name>contextConfigLocation</param-name>
                <param-value>/WEB-INF/dispatcher-servlet.xml</param-
value>
            </init-param>
            <load-on-startup>1</load-on-startup>
            <async-supported>true</async-supported>
        </servlet>
        <servlet-mapping>
            <servlet-name>springMvc</servlet-name>
```

```
            <url-pattern>/</url-pattern>
        </servlet-mapping>
    <!--连接池启用 Web 监控统计功能 -->
        <filter>
            <filter-name>DruidWebStatFilter</filter-name>
            <filter-class>com. alibaba. druid. support. http. WebStatFilter</filter
-class>
            <init-param>
                <param-name>exclusions</param-name>
                <param-value>*. js, *. gif, *. jpg, *. png, *. css, *. ico,/
druid/*</param-value>
            </init-param>
        </filter>
        <filter-mapping>
            <filter-name>DruidWebStatFilter</filter-name>
            <url-pattern>/*</url-pattern>
        </filter-mapping>
        <servlet>
            <servlet-name>DruidStatView</servlet-name>
            <servlet-class>com. alibaba. druid. support. http. StatViewServlet</
servlet-class>
        </servlet>
        <servlet-mapping>
            <servlet-name>DruidStatView</servlet-name>
            <url-pattern>/druid/*</url-pattern>
        </servlet-mapping>
        <!--设置请求编码格式 -->
        <filter>
            <filter-name>encodingFilter</filter-name>
    <filter-class>
    org. springframework. web. filter. CharacterEncodingFilter
    </filter-class>
            <init-param>
                <param-name>encoding</param-name>
                <param-value>UTF-8</param-value>
```

```
        </init-param>
    </filter>
    <filter-mapping>
        <filter-name>encodingFilter</filter-name>
        <url-pattern>/*</url-pattern>
    </filter-mapping>
  <welcome-file-list>
    <welcome-file>/WEB-INF/login.jsp</welcome-file>
  </welcome-file-list>
</web-app>
```

因为启动了数据库连接池监控，所以可以通过 druid/* 的请求映射访问，例如访问 url 为 http://localhost/p11_1/druid/，如图 11－3 所示。

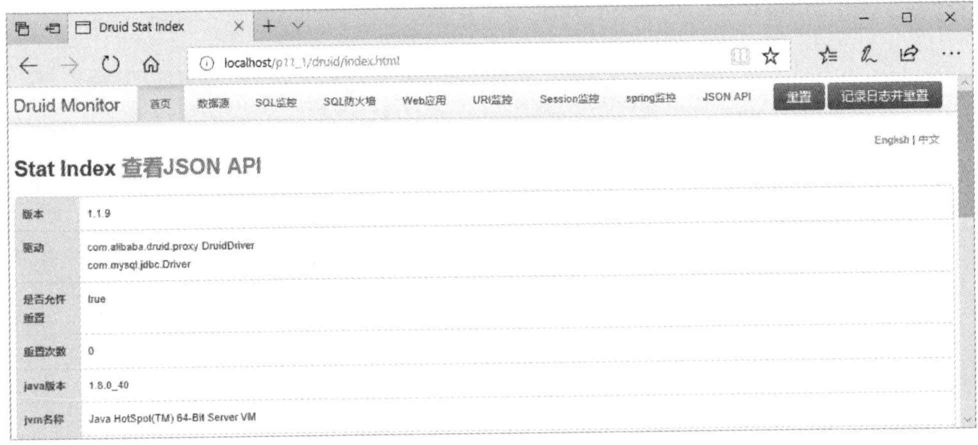

图 11－3　Druid Stat 监控

(7)在工程中新建 CSS 文件和 JSP 页面。参照 p10_1 在工程的 WebContent 下导入样式文件 css/tb.css。然后，在 WEB－INF 下建立 login.jsp 文件，页面中需要 2 个 button，分别用于登录和注册，用于不同请求，其实现代码如下所示。

```
<! -- login.jsp -->
……
<input type="submit" value="登录"
id="login" onclick="myform.action='login.do';myForm.submit()"/>
<input type=" submit " value="注册"
id="regist" onclick="myform.action='regist.do';myForm.submit()"/>
……
```

在 WEB－INF 下建立 error.jsp 文件，当登录或注册失败后将转到该页面，并 5 秒自动转至登录 login.jsp 页面。主要代码如下所示。

<! —— error. jsp ——>

……

登录或注册失败！请点击<a href="newlogin. do">这里</a>重新登录！或者 5 秒后将自动跳转到登录页面！<% response. setHeader (" refresh "," 5; URL = newlogin. do ")；%>

……

在 WEB—INF 下建立 student/query. jsp 文件，用于显示学生信息。代码参考 p10_1 中的 queryStu. jsp，将源代码中的 Struts 2 标签去掉即可，这里不再赘述。

(8)工程运行结果如图 11—4 所示。

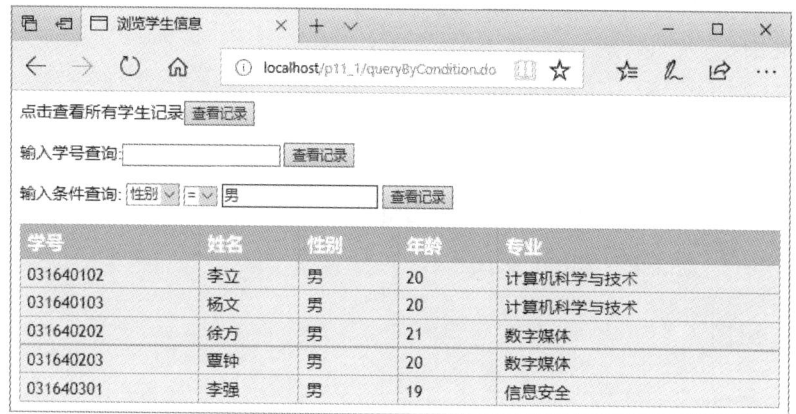

图 11—4　程序运行结果

四、实验思考

1. 按步骤完成实验内容，每个步骤截图保存，形成实验报告。

2. 体会 Struts 2 与 Spring MVC 4 在流程控制逻辑上处理方式的异同，总结各自特点。

3. 实验中通过 Spring 配置了数据库连接池，请尝试通过 Spring 配置 MyBatis 实现数据库连接。

第四部分　综合应用开发

实验十二　Spring 4 开发

一、实验目的

1. 了解 Spring 容器的 7 大模块。
2. 掌握 Spring 开发环境的搭建。
3. 理解并掌握 Spring 中的 AOP、IoC 和 DI 的概念。
4. 掌握 Spring 容器的核心开发技术。

二、基础知识

1. Spring 框架。

简单来讲，Spring 是一个轻量级容器框架。其内部包含 7 大模块，如图 12—1 所示。在 Spring 框架下实现多个子模块的组合，这些子模块之间可以彼此独立，也可以使用其他框架方案加以代替，Spring 希望为企业应用提供一站式的解决方案。

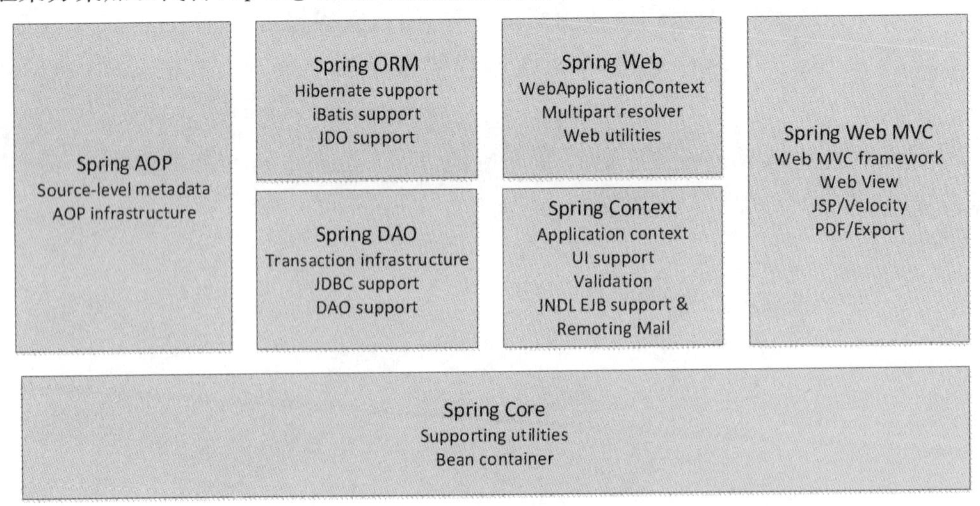

图 12—1　Spring 框架的 7 大模块

Spring 的核心思想包括 IoC(控制反转)、DI(依赖注入)和 AOP(面向切面编程)。

(1)IoC(Inversion of Control，控制反转)的目的是使程序松耦合。可以不必像以往

主动地去新建(New)一个对象,而是在程序运行时依据 Spring 的配置文件动态地创建和调用对象,即把控制权转交给 Spring 容器,由容器根据配置文件去创建实例 bean,并构建各个实例 bean 之间的依赖关系。

(2)IoC 可以根据配置文件动态创建一个对象,如果该对象的调用依赖于其他对象时,就需要通过 DI(Dependency Injection,依赖注入)来实现了。例如,如果要创建 Student 类,发现 Student 类有个属性是 Score 类,而同时 Spring 的 Spring context(应用上下文)组件,统一创建和维护各种 Bean,包括 Bean 之间的依赖关系,而 Spring 对于类之间的依赖关系,会自动进行逐级处理,所有的依赖项都是通过属性注入的方式进行解决。也就是为了创建 Student 类,会自动先创建 Score 类的对象,并以属性注入到 Stduent 类中。这种使用将需要的值注入属性的做法来完成类的完整地创建,就是依赖注入。

(3)AOP(Aspect Oriented Program,面向切面编程)在程序运行时或者在代码编译时、类加载时,动态地将代码切入到指定的类方法以及指定位置上的编程思想,就是面向切面的编程。其中 Spring 中的 AOP 思想是在程序运行时进行的切面编程,也可以像 AspectJ 那样在代码编译时、类加载时实施切面编程。

2. IoC 容器创建 bean。

传统程序通常采用 New 方法和反射机制实例化对象,但 Spring 在 IoC 容器采用配置文件和反射机制来实例化 bean。Spring IOC 实例化 Bean 的方法有如下 3 种。

(1)调用构造器创建 bean 实例。

(2)调用静态工厂方法创建 bean 实例。

(3)调用实例工厂方法创建 bean 实例。

3. AOP 术语。

(1)连接点(Jointpoint):表示需要在程序中插入横切关注点的扩展点,Spring 只支持方法作为连接点。

(2)切入点(Pointcut):选择一组相关连接点的模式,即可以认为连接点的集合。

(3)通知/增强(Advice):是在拦截连接点后执行的行为,通知同时提供了在 AOP 中需要在切入点进行的手段。包括前置通知(before advice)、后置通知(after advice)、环绕通知(around advice)、异常通知(AfterThrowing advice)、返回通知(fterRunning advice)。

(4)方面/切面(Aspect):横切关注点的模块化,可以认为是通知、引入和切入点的组合。

(5)引入(Introductions):也称为内部类型声明,为已有的类添加额外新的字段或方法,Spring 允许引入新的接口(必须对应一个实现)到所有被代理对象(目标对象)。

(6)目标对象(Target Object):需要被织入横切关注点的对象,即该对象是切入点选择的对象,需要被通知的对象。由于 Spring AOP 通过代理模式实现,这个对象永远是被代理对象。

(7)织入(Weaving):织入是一个过程,是将切面应用到目标对象从而创建出 AOP 代理对象的过程,织入可以在编译期、类装载期、运行期进行。

三、实验步骤

1. 下载 Spring 开发 jar 包。

(1)在 http://maven. springframework. org/release/org/springframework/spring/ 网站下载最新 Spring 框架开发 jar 包。如本书实验所下载的 spring—framework—5. 0. 5. RELEASE—dist. zip。

(2)将下载的压缩包解压,得到的文件中的 libs 目录下就是 Spring5 框架依赖的 jar 包文件。其中核心 jar 包见表 12—1。

表 12—1　Spring5 的核心 jar 包文件

序号	包名
1	spring—beans—5. 0. 5. RELEASE. jar
2	spring—context—5. 0. 5. RELEASE. jar
3	spring—core—5. 0. 5. RELEASE. jar
4	spring—expression—5. 0. 5. RELEASE. jar

2. 在 Eclipse 中添加 Spring 开发插件。给 Eclipse 添加 Spring tools 插件,具体方法略。通过插件可以方便地给工程新建 Spring 的配置文件。

3. 新建一个 Java 工程,实现 Spring5 中的 IoC(控制反转)和 DI(依赖注入)编程。

(1)在 Eclipse 中,新建一个 Java 工程 p12_1,在工程中新建 lib 目录,并导入 Spring5 的核心开发包以及其他相关 jar 包,然后通过 Configure build path 把这些 jar 包引入到项目库文件中,如图 12—2 所示。

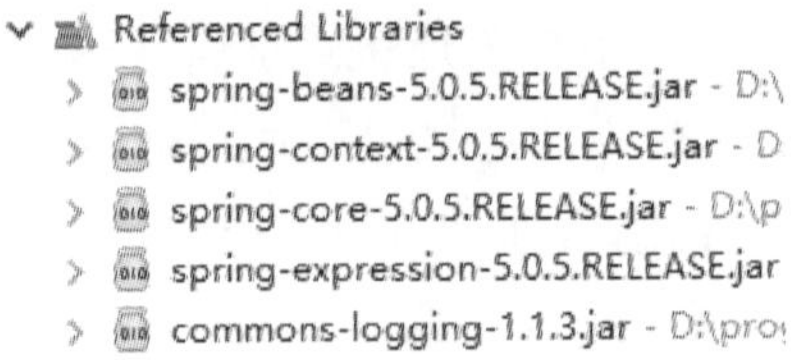

Referenced Libraries
spring-beans-5.0.5.RELEASE.jar - D:\
spring-context-5.0.5.RELEASE.jar - D
spring-core-5.0.5.RELEASE.jar - D:\p
spring-expression-5.0.5.RELEASE.jar
commons-logging-1.1.3.jar - D:\pro

图 12—2　工程引入的库文件

(2)在 p12_1 中新建 cn. hbmy. p12_1. bean 包,在包中新建 Student 类,部分代码如下所示。

```
//Student. java
……
public class Student {
private String stuNum;
```

```
    private String stuName;
    private String stuSex;
    private int stuAge;
    private String stuMajor;
    public Student(String stuNum,String stuName,String stuSex,int stuAge,String stuMajor){
    this.stuNum=stuNum;
    this.stuName=stuName;
    this.stuSex=stuSex;
    this.stuAge=stuAge;
    this.stuMajor=stuMajor;
    }
    public String toString( ){
    return ("学号:"+stuNum+" 姓名:"+stuName+" 性别:"+stuSex+" 年龄:"+stuAge+" 专业:"+stuMajor);
    }
    //省略 setter 和 getter 方法
    ……
```

注意,这里定义构造方法的目的是通过构造方法依赖注入。

(3)在工程中新建 Spring 的配置文件。右击 src 目录,在弹出的快捷菜单中选择 New | Other,在弹出的对话框中选择 Spring | Spring Bean Configuration File,点击 Next 按钮,填写配置文件名为 applicationContext.xml,点击 Finish 按钮完成配置文件的创建,该配置文件代码如下所示。

```
<? xml version="1.0" encoding="UTF-8"? >
<beans xmlns="http://www.springframework.org/schema/beans"
xmlns:xsi="http://www.w3.org/2001/XMLSchema-instance"
xsi:schemaLocation="http://www.springframework.org/schema/beans http://www.springframework.org/schema/beans/spring-beans.xsd">
<bean id="stu1" class="cn.hbmy.p12_1.bean.Student">
<constructor-arg index="0" type="java.lang.String">
<value>031740101</value>
</constructor-arg>
<constructor-arg index="1" type="java.lang.String">
<value>李星星</value>
</constructor-arg>
```

```
<constructor-arg index="2" type="java.lang.String">
<value>男</value>
</constructor-arg>
<constructor-arg index="3" type="int">
<value>19</value>
</constructor-arg>
<constructor-arg index="4" type="java.lang.String">
<value>信息安全</value>
</constructor-arg>
</bean>
</beans>
```

注意，虽然配置文件名可自定义，但我们建议 Spring 容器的配置文件取名 applicationContext.xml，并放置在 src 目录下，以方便访问。另外，本例是通过配置文件实现构造器注入方式创建了 Student 实例。

(4)在工程中新建 cn.hbmy.p12_1.app 包，并在包中新建测试类 SpringEnvTest，对构造器注入方式进行测试，SpringEnvTest 类的部分代码如下所示。

```
//SpringEnvTest.java
……
public class SpringEnvTest {
private static ApplicationContext ctx;
public static void main(String[] args) {
ctx=new ClassPathXmlApplicationContext("applicationContext.xml");
Student stu=(Student) ctx.getBean("stu1");
System.out.println(stu.toString());
}
}
```

这里读取 applicationcontext.xml 文件的方式是利用 ClassPathXmlApplicationContext，这种方式配置文件应该放在与类包相同的路径下。运行 Java 程序，从控制台得到运行结果，从而体会有关构造方法的依赖注入原理。

(5)在 p12_1 中，新建 Score 类，Score 类的部分代码如下所示。

```
//score.java
package cn.hbmy.p12_1.bean;
public class Score {
private String couId;
private int couSc;
```

```
public String toString( ) {
return (" 课程号:"+couId+" 分数:"+couSc);
}
//省略 setter 和 getter 方法
……
```

注意,这个的 Score 类并没有创建构造方法,可见新建 Score 类的对象,并不再使用构造器依赖注入方式了。下面步骤将实现通过设置注入的方式创建 Score 实例。

(6)修改 Spring 的配置文件 applicationContext.xml,通过设置注入的方式实现 Score 类的实例化,增加部分的代码如下所示。

```
……
<bean id="score1" class="cn.hbmy.p12_1.bean.Score">
<property name="couId">
<value>003</value>
</property>
<property name="couSc">
<value>031740102</value>
</property>
</bean>
<bean id="score2" class="cn.hbmy.p12_1.bean.Score">
<property name="couId">
<value>002</value>
</property>
<property name=" couSc ">
<value>76</value>
</property>
</bean>
……
```

(7)修改测试类 SpringEnvTest,在 main 方法中新增如下所示代码。

```
Score score=(Score)ctx.getBean("score1");
System.out.println(score.toString( ));
score=(Score)ctx.getBean("score2");
System.out.println(score.toString( ));
```

运行程序,在控制台查看运行结果,体会此例在依赖注入方式上的不同。

(8)修改 Student 类,在原有属性基础上增加类属性 scores 并修改构造方法和 toString 方法,具体代码如下所示。

```
//Student. java
……
private List<Score> scores;
public Student(String stuNum, String stuName, String stuSex, int stuAge, String stuMajor, List<Score> scores){
this. stuNum=stuNum;
this. stuName=stuName;
this. stuSex=stuSex;
this. stuAge=stuAge;
this. stuMajor=stuMajor;
this. scores=scores;
}
public String toString( ){
String strSc="";
for(Score s:scores)
{
strSc=strSc+s. toString( )+",";
}
return("学号:"+stuNum+" 姓名:"+stuName+" 性别:"+stuSex+" 年龄:"+stuAge+" 专业:"+stuMajor+strSc. substring(0, strSc. length( )-1));
}
//省略 setter 和 getter 方法
……
```

(9)再次修改配置文件 applicationContext. xml,在其中增加对 Student 实例的配置,通过设置属性注入 scores 的值,增加的代码如下所示。

```
……
<constructor-arg index="5" name="scores">
<list>
<ref bean="score1"/>
<ref bean="score2"/>
</list>
</constructor-arg>
……
```

运行 Java 工程,在控制台记录输出结果,体会 List 属性的注入方式。

(10)在工程的 cn. hbmy. p12_1. bean 包中,新增 StudentFactory 类,通过调用静态

工厂方法创建 bean。StudentFactory 类代码如下所示。

```
//StudentFactory.java
package cn.hbmy.p12_1.bean;
import java.util.List;
public class StudentFactory {
public static Student getInstance(String stuNum,String stuName,String stuSex,
int stuAge,String stuMajor,List<Score> scores) {
System.out.println("调用静态工厂创建 bean……");
return new Student(stuNum,stuName,stuSex,stuAge,stuMajor,scores);
}
}
```

(11)在工程的 cn.hbmy.p12_1.bean 包中，新增 StudentInstanceFactory 类，通过调用实例工厂方法创建 bean。StudentInstanceFactory 类代码如下所示。

```
//StudentInstanceFactory.java
package cn.hbmy.p12_1.bean;
public class StudentInstanceFactory {
public Student getInstance(String stuNum, String stuName, String stuSex, int
stuAge,String stuMajor,List<Score> scores) {
System.out.println("调用实例工厂创建 bean……");
return new Student(stuNum,stuName,stuSex,stuAge,stuMajor,scores);
}
}
```

(12)修改配置文件 applicationContext.xml，实现调用静态工厂和实例工厂的方法创建 bean。在配置文件中增加如下代码。

```
……
<!--定义静态工厂-->
<bean id="studentfactory"
class="cn.hbmy.p12_1.bean.StudentFactory" factory-method="getInstance">
<constructor-arg index="0" value="0317401"></constructor-arg>
<constructor-arg index="1" value="简娜"></constructor-arg>
<constructor-arg index="2" value="女"></constructor-arg>
<constructor-arg index="3" value="19"></constructor-arg>
<constructor-arg index="4" value="数字媒体"></constructor-arg>
<constructor-arg index="5">
<list>
```

```
<ref bean="score1" />
<ref bean="score2" />
</list>
</constructor-arg>
</bean>
<! --定义实例工厂 -->
<bean id="studentinstancefactory"
class="cn.hbmy.p12_1.bean.StudentInstanceFactory"></bean>
<! --利用实例工厂创建 bean -->
<bean id="studentinstance" factory-bean="studentinstancefactory"
factory-method="getInstance">
<constructor-arg index="0" value="0317401"></constructor-arg>
<constructor-arg index="1" value="张鹏"></constructor-arg>
<constructor-arg index="2" value="男"></constructor-arg>
<constructor-arg index="3" value="20"></constructor-arg>
<constructor-arg index="4" value="数字媒体"></constructor-arg>
<constructor-arg index="5">
<list>
<ref bean="score1" />
<ref bean="score2" />
</list>
</constructor-arg>
</bean>
......
```

(13)修改测试类 SpringEnvTest。在测试类中尝试调用静态工厂和实例工厂方法创建 bean 实例。部分代码如下所示。

```
//SpringEnvTest.java
......
public static void main(String[] args) {
ctx=new ClassPathXmlApplicationContext("applicationContext.xml");
Student stu=(Student)ctx.getBean("stu1");
System.out.println(stu.toString());
Score score1=(Score)ctx.getBean("score1");
System.out.println(score1.toString());
Score score2=(Score)ctx.getBean("score2");
```

```
System.out.println(score2.toString());
Student sf=(Student)ctx.getBean("studentfactory");
System.out.println(sf.toString());
Student sif=(Student)ctx.getBean("studentinstance");
System.out.println(sif.toString());
}
}
```

程序分别用构造器注入、属性注入、静态工厂注入和实例工厂注入的方法分别创建了 bean 实例 stu、score1 和 score2、sf、sif 等 4 组对象。运行程序，记录控制台输出结果，并对结果进行分析。

4. 新建一个 Java 工程，实现 Spring5 中的 AOP(面向切面编程)。

(1)在 Eclipse 中，新建一个 Java 工程 p12_2，在工程中新建 lib 目录，并导入 Spring5 的核心开发包以及其他相关 jar 包，然后通过 Configure build path 把这些 jar 包引入到工程库文件中，如图 12—3 所示。

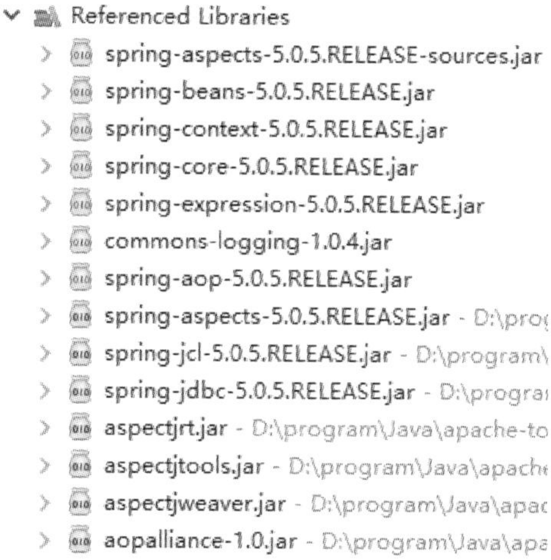

```
Referenced Libraries
  spring-aspects-5.0.5.RELEASE-sources.jar
  spring-beans-5.0.5.RELEASE.jar
  spring-context-5.0.5.RELEASE.jar
  spring-core-5.0.5.RELEASE.jar
  spring-expression-5.0.5.RELEASE.jar
  commons-logging-1.0.4.jar
  spring-aop-5.0.5.RELEASE.jar
  spring-aspects-5.0.5.RELEASE.jar - D:\pro
  spring-jcl-5.0.5.RELEASE.jar - D:\program\
  spring-jdbc-5.0.5.RELEASE.jar - D:\progra
  aspectjrt.jar - D:\program\Java\apache-to
  aspectjtools.jar - D:\program\Java\apach
  aspectjweaver.jar - D:\program\Java\apa
  aopalliance-1.0.jar - D:\program\Java\ap
```

图 12—3 工程库文件

其中除了 Spring 的开发包外，还需要另行下载 aspectj－1.9.0.jar，并添加其中的 aspectjrt.jar、aspectjtools.jar 和 aspectjweaver.jar 到工程中。然后下载 aopalliance－1.0.jar，并添加到工程中。

(2)新建 cn.hbmy.p12_2.service 包，并在包中新建 UserService 接口，代码如下所示。

```
//UserService.java
package cn.hbmy.p12_2.service;
public interface UserService {
public void login();
public void regist();
```

```
}
```

(3)新建 cn. hbmy. p12_2. service. impl 包,并在包中新建 UserService 接口方法的实现类 UserServiceImpl 类,该类定义的方法将作为切入点进行切面编程。代码如下所示。

```
//UserServiceImpl. java
package cn. hbmy. p12_2. service. impl;
import cn. hbmy. p12_2. service. UserService;
public class UserServiceImpl implements UserService {
public void login( ) {
System. out. println("用户登录……");
}
public void regist( ) {
System. out. println("用户注册……");
}
}
```

(4)新建 cn. hbmy. p12_2. aspect 包,并新增 SecurityAspect 类,作为切入类,其中的 beforeAdvice 方法将作为前置通知。代码如下所示。

```
//SecurityAspect. java
package cn. hbmy. p12_2. aspect;
public class SecurityAspect {
public void beforeAdvice( ) {
        System. out. println("用户权限检验……");
    }
}
```

(5)在 src 目录下新建 applicationContext. xml 文件,文件代码如下所示。

```
<! --applicationContext. xml-->
<? xml version="1. 0" encoding="UTF-8"? >
<beans xmlns="http://www. springframework. org/schema/beans"
xmlns:xsi="http://www. w3. org/2001/XMLSchema-instance"
xmlns:aop="http://www. springframework. org/schema/aop"
xmlns:lang="http://www. springframework. org/schema/lang"
xsi:schemaLocation="http://www. springframework. org/schema/beans http://
www. springframework. org/schema/beans/spring-beans. xsd
http://www. springframework. org/schema/aop http://www. springframework.
org/schema/aop/spring-aop-4. 3. xsd
```

```
http://www.springframework.org/schema/context http://www.springframework.org/schema/context/spring-context-4.3.xsd">
<aop:aspectj-autoproxy  proxy-target-class="true"/>
<!--被切入类 -->
<bean id="userServiceImpl" class="cn.hbmy.p12_2.service.impl.UserServiceImpl"/>
<!--切入的类 -->
<bean id="aspect" class="cn.hbmy.p12_2.aspect.SecurityAspect"/>
<aop:config>
<!-- 定义被切入的位置-->
<aop:pointcut id="security"
expression="execution( * cn.hbmy.p12_2.service.impl.*.*(..))"/>
<!--aspect 设置切面支持类 -->
<aop:aspect ref="aspect">
<!-- method 用来引用切面通知实现类中的方法 -->
<aop:before pointcut-ref="security" method="beforeAdvice"/>
</aop:aspect>
</aop:config>
</beans>
```

注意，代码“execution(* cn. hbmy. p12_2. service. impl. *. *(..))”表示设置切入点，第 1 个 * 代表所有类型的返回值，第 2 个 * 是代表 cn. hbmy. p12_2. service. impl 包下的所有类，第三个是类下的所有方法，括号中两个点表示任意个形参。另外注意，代码<aop:aspectj-autoproxy proxy-target-class="true"/>可以防止出现代理异常。

(6)新建 cn. hbmy. p12_2. app 包，并新建 SpringEnvTest 测试类，类代码如下所示。

```
//SpringEnvTest.java
public class SpringEnvTest {
private static ApplicationContext ctx;
public static void main(String[] args) {
ctx=new ClassPathXmlApplicationContext("applicationContext.xml");
UserService userService = ctx.getBean("userServiceImpl", UserServiceImpl.class);
userService.login();
}
}
```

运行程序并记录控制台输出结果，体会切面编程思想。

四、实验思考

1. 按步骤完成实验内容，每个步骤截图保存，形成实验报告。

2. 实验中读取配置文件 application Context. xml 的方法是通过 Class Path Xml Application Context，列举出其他几种读取配置文件的方法，并说明其特点。

3. 通过实验体会 AOP 的编程思想，并举例 AOP 有哪些应用场景。

实验十三　Spring+ Spring MVC+ MyBatis 应用开发

一、实验目的

1. 掌握 Spring、Spring MVC 和 MyBatis 整合开发 Web 应用的方法。
2. 掌握 Spring 拦截器编程实现用户登录验证的方法。
3. 掌握 Ajax 实现页面输入验证的方法。
4. 掌握 Spring AOP+ Log4j2 编程实现日志操作。
5. 掌握 SiteMesh 框架实现网页布局和修饰。

二、基础知识

1. SSM 开发框架。

SSM 开发框架指的是 Spring+Spring MVC+MyBatis 的整合，常用于较轻量级的 Web 开发应用。本实验中通过整合这 3 个框架，实现 Web 项目的开发。其中，运用 Spring AOP 实现了项目日志管理；运用 Spring MVC 的拦截器编程实现用户登陆验证；运用 Spring MVC 作为整体 Web 项目的开发框架；运用 MyBatis 实现数据持久层操作。

2. Spring MVC 拦截器。

Spring MVC 中的 Interceptor 主要用于拦截用户请求并作出相应处理。Interceptor 的定义通过 2 种方式实现，第 1 种方式是直接实现 Handler Interceptor 接口，或者是继承实现了该接口的类，比如继承 Handler Interceptor Adapter 抽象类。第 2 种方式是实现 Spring 的 Web Request Interceptor 接口，或者是继承实现了该接口的类。

3. Log4j2 基础。

log4j2 是 Apache 的一个开放源代码的项目，通过使用 log4j2，我们可以控制日志信息输出。比如，可以输出到控制台、文件、GUI 组件、套接口服务器、NT 的事件记录器、UNIX Syslog 守护进程等。我们还可以控制每一条日志的输出格式，以及通过定义每一条日志信息的级别，我们能够更加细致地控制日志的生成过程。

Log4j2 有三个重要组成部分：日志记录(Logger)、输出端(Appender)和日志格式化(Layout)。

(1)Logger:控制要启用或禁用哪些日志记录语句,并对日志信息进行级别限制。

(2)Appender :指定了日志将打印到控制台还是文件中。

(3)Layout :控制日志信息的显示格式。

Log4j 将要输出的 Log 信息定义了 8 种级别,按照从低到高排列为:All < Trace < Debug < Info < Warn < Error < Fatal < OFF,当输出时,只有级别高过配置中规定级别的信息才能真正输出,这样就可以方便地配置不同情况下要输出的内容,而不需要更改代码。log4j2 的配置文件只能采用. xml, . json 或者 . jsn 格式。在默认情况下,系统选择配置文件的优先顺序如下所示。

(1)src 目录下,名为 log4j－test. json 或者 log4j－test. jsn 的文件。

(2)src 目录下,名为 log4j2－test. xml 的文件。

(3)src 目录下,名为 log4j. json 或者 log4j. jsn 文件。

(4)src 目录下,名为 log4j2. xml 的文件。

并且,log4j2 在集成到 Servlet 3 以上版本的 Web 项目时做了简化,仅需要导入 log4j－api、log4j－core 和 log4j－web 这 3 个 jar 包,然后配置好 log4j2. xml 文件即可使用。也就是不需要在 web. xml 文件中进行任何配置就可应用。

4. Ajax 基础。

Ajax 采用异步方式实现交互式网页应用,即可动态更新部分网页数据,而无需加载整个网页。其原理是通过 JavaScript 向服务器提出 XMLHTTPRequest 请求,使得在不重载页面的情况与 Web 服务器交换数据。

5. SiteMesh 基础。

SiteMesh 是一个网页布局和修饰的框架,用于分离网页内容和结构。SiteMesh 采用过滤器方式,对预备返回给客户端浏览器的已处理页面进行装饰,以此达到网页的内容和页面结构分离的目的。使用 SiteMesh 框架,在项目中能够帮助创建一致的页面布局和外观,比如,一致的导航条、banner、版权信息等。

三、实验步骤

实验将整合 Spring、Spring MVC 和 MyBatis 框架开发 Web 数据库访问应用。将使用 Spring MVC 框架实现 Web 项目的整体开发;通过 Spring AOP 实现项目的日志管理;运用 Spring MVC 的拦截器编程实现用户登陆验证;通过 MyBatis 框架实现数据持久层的操作;通过 Ajax 实现页面输入验证;通过 Spring AOP+ Log4j2 编程实现日志操作;通过 SiteMesh 框架实现网页布局和修饰。实验中所有文件的完整代码收录在附录中。

1. 新建一个 Web 工程,搭建开发环境。

(1)下载工程需要的相关开发包。通过互联网分别获取支持这 3 个框架的相关开发包。之前的实验已说明获取相应开发包的方式,这里不再赘述。本实验需要的开发包见表 13－1。

表 13—1　实验需要的相关 jar 包文件

序号	包名	序号	包名
1	spring—beans—5. 0. 5. RELEASE. jar	2	spring—context—5. 0. 5. RELEASE. jar
3	spring—core—5. 0. 5. RELEASE. jar	4	spring—web—5. 0. 5. RELEASE. jar
5	spring—webmvc—5. 0. 5. RELEASE. jar	6	spring—aop—5. 0. 5. RELEASE. jar
7	spring—expression—5. 0. 5. RELEASE. jar	8	spring—tx—5. 0. 5. RELEASE. jar
9	spring—aop—5. 0. 5. RELEASE. jar	10	spring—jdbc—5. 0. 5. RELEASE. jar
11	aopalliance—1. 0. jar	12	aspectjrt. jar
13	aspectj tools. jar	14	aspectjweaver. jar
15	commons—logging—1. 0. 4. jar	16	druid—1. 1. 9. jar
17	jackson—annotations—2. 9. 5. jar	18	jackson—core—2. 9. 5. jar
19	jackson—databind—2. 9. 5. jar	20	jsqlparser—0. 9. 5. jar
21	jstl. jar	22	log4j—api—2. 11. 0. jar
23	log4j—core—2. 11. 0. jar	24	log4j—web—2. 11. 0. jar
25	mybatis—3. 4. 6. jar	26	mybatis—spring—1. 3. 2. jar
27	mysql—connector—java—5. 1. 7—bin. jar	28	org. json. jar
29	pagehelper—5. 1. 3. jar	30	pager—taglib. jar
31	sitemesh—3. 0. 1. jar	32	standard. jar

(2)在 Eclipse 中,新建一个 Web 工程 p13,在工程中导入之前准备好的开发 jar 包到项目的 WEB—INF/lib 目录下。

2. 创建数据库,并根据数据表抽象实体类。之后,这里创建的实体类和数据表将通过 MyBatis 框架构建 ORM(对象关系映射)。

(1)这里采用之前实验所使用的 edu_db 数据库,将使用其中的表 edu_user 和表 edu_stu,其创建过程不再赘述。

(2)根据数据表创建实体类。新建 cn. hbmy. p13. entity 包,在包中创建 Student 和 User 类,分别对应表 edu_user 和表 edu_stu。在实验八中使用 Eclipse 的 MyBatis 插件实现了根据配置文件自动生成实体类,这里通过自定义完成实体类的创建。

3. 通过 Spring MVC 框架实现流程控制逻辑、业务逻辑和数据访问逻辑的分离,通过 MyBatis 框架实现 DAO 数据访问,同时在业务层使用 MyBatis 的分页插件 PageHelper 实现分页查询。

(1)通过 MyBatis 框架创建 DAO 数据访问层。新建 cn. hbmy. p13. mapper 包,在包中创建 StudentMapper 接口和 UserMapper 接口,分别用于封装不同数据对象的所有原子操作。继续在该包下新建 StudentMapper. xml 和 UserMapper. xml 映射文件,Mapper 映射文件是对接口的具体实现,是 MyBatis 框架的核心内容。在映射文件中通

过 id 属性来设定 SQL 执行语句与接口方法的关联，通过 MyBatis 框架实现了数据库的底层访问逻辑，关于 MyBatis 框架的配置将在后面详述。这种方式与 Struts 实验中的 DAO 层的实现方式不同，需要读者认真体会。

（2）创建 Service 业务逻辑层。新建 cn. hbmy. p13. service 包，在包中创建 IStudentService 接口和 IUserService 接口，以及 StudentService 和 UserService 这 2 个相应的接口实现类。Service 层和 Struts 实验中意义相同，都是用于封装业务逻辑，然后在业务逻辑 Service 中调用 DAO 层，从而实现了业务逻辑和数据访问逻辑的解耦。

注意，在 Service 层使用分页插件，实现分页查询的业务逻辑。要使用该插件需要导入插件工具包，实验中使用的是 pagehelper—5. 1. 3. jar 和 pager—taglib. jar。然后在分页查询的方法中使用工具包中封装的 PageInfo 类，帮助完成分页查询功能。

（3）创建 Controller 控制层。新建 cn. hbmy. p13. controller 包，在包中创建 IndexController、UserController 和 StudentController 等 3 个控制器。分别用于处理访问页面的请求、访问用户操作的请求和访问学生信息操作的请求。由于本工程中将所有 JSP 页面放置于 WEB—INF 目录下，因此希望通过 IndexController 控制器来控制对 JSP 页面的访问；通过 UserController 控制器响应对用户操作的请求，比如登录、注册、修改用户密码等；通过 StudentController 控制器响应对学生信息操作的请求，比如"增删改查"等操作请求。并且，在 UserController 和 StudentController 中，实现了对前端基于 Ajax 的访问请求处理，对于这些请求通过返回 Json 对象数据给予前端响应。

需要说明的是，在 Spring 中提供了@Component、@Controller、@Service、@Repository 等注解来标注 Spring Bean。@Component 用于标注普通 Spring Bean；@Controller 用于标注控制器组件；@Service 用于标注业务逻辑组件；@Repository 用于标注 DAO 组件。前面对于各个 Spring Bean 也需要使用相应注解，以实现在 Spring 配置文件中对这些 Spring Bean 的自动搜索和装配。实验中，在 applicationContext. xml 文件使用＜context：component－scan base－package＝"cn. hbmy. p13"＞标签后，spring 可以自动去扫描 base—package 下或者子包下所有 java 类文件，并将扫描到的用@Component 、@Controller、@Service、@Repository 等这些注解的类，注册成 Bean。

4. 通过 Spring MVC 拦截器实现用户登录验证。

（1）新建 cn. hbmy. p13. interceptor 包，在包中创建 LoginInterceptor，用于拦截器实现用户的登录验证。这里的拦截器是通过继承 HandlerInterceptorAdapter 抽象类实现，通过重载其 preHandle 方法，实现在请求达到控制器之前被拦截处理。值得注意的是，当设置了拦截器，将拦截所有请求，此时对登录或注册的请求也将进行拦截，因此需要解决不拦截登录和注册请求的问题。解决的方法包括在下一步的配置文件中设置不拦截的资源，或在拦截器设计中通过代码判断放行的资源。实验中使用的方法是后者。

（2）在 Spring MVC 的配置文件 dispatcher—servlet. xml 中配置拦截器。注意在配置拦截器后，默认的 Spring MVC 拦截器会拦截静态资源文件，这也是这种 REST 风格的弊端。实验中使用的解决方法是在配置文件 dispatcher－servlet. xml 中增加了对拦截器不用拦截资源的设置，例如，在<mvc:exclude－mapping path＝"/＊＊/＊. css"/>中，设置了不必拦截所有的 css 资源文件的条件。

5. 通过 Spring AOP 和 log4j2 实现项目日志操作。借鉴 AOP 的思想，将日志操作做成一个切面，切入到需要进行日志操作的业务逻辑或者控制逻辑的前后，这样既利于软件开发阶段排查错误，也有利于软件应用阶段的系统维护。

(1)在 src 目录下创建 log4j2. xml 文件，用于日志配置。Servlet 3 以上版本的 Web 项目在集成 log4j2 时，仅需要导入相关开发 jar 包，然后配置好 log4j2. xml 文件即可使用。实验中的日志设置了 2 个级别的输出，一个是 debug，用于调试阶段在控制台中输出内存参数；一个是 info，用于运行阶段在文件中输出用户所有的操作信息。

(2)在 cn. hbmy. p13. interceptor 包中新建 LogAspect 类，并通过 AOP 注解的方式实现对 Controller 中用户相应方法的切面编程。即当用户调用相应方法时，将切入完成相关日志操作。这里使用注解的方式进行 AOP 编程，与之前实验十二中的 AOP 编程相比较，显得更加简洁，仅需要在 Spring 的配置文件中增加 AOP 的支持，并注册 LogAspect 类。

(3)在实验中的 Spring 配置文件中增加 AOP 支持，并注册 LogAspect 类。当前实验在 Spring MVC 的配置文件 dispatcher－servlet. xml 中进行配置，需要注意在配置文件中要引入 AOP 的名称空间 xmlns:aop，还要在 xsi:schemaLocation 中加入 aop 的 xsd 文件。

6. 使用 SiteMesh 框架实现网页布局和修饰。

(1)下载和安装 SiteMesh。实验中使用的版本是 sitemest－3. 0. 1. jar。

(2)配置 web. xml 文件。在 web. xml 文件配置 SiteMeshFilter 过滤器。SiteMesh 实质上是一个页面过滤器，过滤所有需要使用 SiteMesh 框架的页面文件。

(3)配置 SiteMesh 的配置文件 decoractors. xml。主要用于配置哪些页面需要装载装饰页，哪些不需要。实验中使用<decorator name＝"main" page＝"main. jsp">设置装饰器，并设置需要使用装饰器的目录路径。

(4)创建装饰页面。在 WEB－INF 下创建 decoractors 目录，并在目录下新建 4 个页面，footer. jsp、header. jsp、menu. jsp 和 main. jsp。前 3 个 jsp 页面都是为 main. jsp 服务，分别描述装饰后页面的头部、底部和菜单，main. jsp 即是装饰器。

7. 整合 Spring、Spring MVC 和 MyBatis 框架。之前的实验分别实现了 Spring、Spring MVC 和 MyBatis 框架应用，这里通过配置文件实现 3 个框架的整合。

(1)创建 MyBatis 相关配置文件。实验八的 MyBatis 实验是利用 Eclipse 的 MybBatis 插件，通过 generatorConfig. xml 文件自动生成与数据表对应的实体类和对应的 mapper. xml 映射文件。这里，我们将手动创建相关文件。在 src 目录下，新建 Mybatis－Config. xml 文件。由于在 Spring 中整合了 MyBatis，因此可在 Spring MVC

的配置文件中配置数据源，所以 Mybatis－Config. xml 文件的内容相对较少，实验中只是对实体类的别名进行了设置，以及注册了有哪些需要引入的 mapper. xml 映射文件。

（2）创建 Spring 的相关配置文件。本实验将 Spring 和 Spring MVC 的配置作了简单分离，在 src 目录下创建了 Spring 的配置文件 applicationContext. xml，而在 WEB－INF 下创建了 Spring MVC 的配置文件 dispatcher － servlet. xml。在 applicationContext. xml 中，主要实现数据库的连接，比如 mybatis 的 sqlSessionFactory 等配置，以及事务、扫描和注解的配置，等等；而在 dispatcher－servlet. xml 中则是针对 Web 的配置，比如视图解析器配置、访问静态资源配置、拦截器配置，等等。最后需要在工程的 web. xml 文件中加载配置文件。

8. 创建前端的 JSP 页面。在工程的 WEB－INF 目录下新建 jsp 目录，将所有的 JSP 文件放置在该目录。

（1）在 jsp 目录下新建 common 目录。在该目录下新建 login. jsp 页面用于用户登录；新建 resources. jsp 页面用于统一装载静态资源文件；新建 welcome. jsp 页面用于登陆后的欢迎页面。

（2）在 jsp 目录下新建 student 目录，用于存放所有针对学生信息的操作页面。新建 create. jsp 页面用于新增学生信息；新建 delete. jsp 用于删除学生信息；新建 query. jsp 用于查询学生信息；新建 update. jsp 页面用于更新学生信息；新建 updatelist. jsp 页面用于查询需要更新的学生信息。

（3）在 jsp 目录下新建 user 目录，用于存放所有针对用户的操作页面。新建 register. jsp 页面用于用户注册，新建 updatepwd. jsp 页面用于修改用户密码。

需要注意的是，当 JSP 页面中使用了 Ajax 时，需要引入相关 js 文件。

四、实验思考

1. 按步骤完成实验内容，每个步骤截图保存，形成实验报告。

2. 思考比较实验中 Mybatis 配置 Mapper 的方法与整合了 Spring MVC 后配置 Mapper 的方法有什么不同，并举出 Mybatis 整合 Spring MVC 的方式有哪几种。

3. 体会用户权限验证的 Struts2 拦截器实现和 Spring AOP 实现的异同。

附录　实验十三详细代码

一、配置文件

1. web. xml.

```
<? xml version="1.0" encoding="UTF-8"? >
<web-app xmlns:xsi="http://www.w3.org/2001/XMLSchema-instance"
xmlns="http://java.sun.com/xml/ns/javaee"
xmlns:web="http://java.sun.com/xml/ns/javaee/web-app_2_5.xsd"
xsi:schemaLocation="http://java.sun.com/xml/ns/javaee http://java.sun.com/xml/ns/javaee/web-app_2_5.xsd"id="WebApp_ID" version="2.5">
<display-name>p13</display-name>

<! --声明应用范围内的上下文初始化参数,Spring 的监听器可以通过这个上下文参数来获取 contextConfigLocation.xml 的位置。-->
<context-param>
<param-name>contextConfigLocation</param-name>
<param-value>
classpath:applicationContext.xml
</param-value>
</context-param>

<! --创建 Spring 的监听器 -->
<listener>
<listener-class>org.springframework.web.context.ContextLoaderListener</listener-class>
</listener>

<! --在 DispatcherServlet 的初始化过程中,框架会在 WEB-INF 文件夹下寻找名为 *-servlet.xml 的配置文件,生成文件中定义的 bean。-->
<servlet>
```

```
<servlet-name>DispatcherServlet</servlet-name>
<servlet-class>org. springframework. web. servlet. DispatcherServlet</servlet
-class>
<init-param>
<param-name>contextConfigLocation</param-name>
<param-value>/WEB-INF/dispatcher-servlet. xml</param-value>
</init-param>
<load-on-startup>1</load-on-startup>
</servlet>
<servlet-mapping>
<servlet-name>DispatcherServlet</servlet-name>
<url-pattern>/</url-pattern>
</servlet-mapping>

<! -- sitemesh 过滤器定义 -->
<filter>
<filter-name>sitemesh</filter-name>
<filter-class>com. opensymphony. sitemesh. webapp. SiteMeshFilter</filter-
class>
</filter>
<filter-mapping>
<filter-name>sitemesh</filter-name>
<url-pattern>/ * </url-pattern>
</filter-mapping>

<! --全局过滤,字符编码 -->
<filter>
<filter-name>CharacterFilter</filter-name>
<filter-class>org. springframework. web. filter. CharacterEncodingFilter</
filter-class>
<init-param>
<param-name>encoding</param-name>
<param-value>UTF-8</param-value>
</init-param>
</filter>
```

```
<filter-mapping>
<filter-name>CharacterFilter</filter-name>
<url-pattern>/*</url-pattern>
</filter-mapping>

<welcome-file-list>
<welcome-file>/WEB-INF/jsp/common/login.jsp</welcome-file>
</welcome-file-list>
</web-app>
```

2. jdbc. properties.

```
driver=com.mysql.jdbc.Driver
url=jdbc:mysql://localhost:3306/edu_db?characterEncoding=utf8
username=root
password=123456
```

3. applicationContext. xml.

```
<?xml version="1.0" encoding="UTF-8"?>
<beans xmlns="http://www.springframework.org/schema/beans"
xmlns:xsi="http://www.w3.org/2001/XMLSchema-instance"
xmlns:context="http://www.springframework.org/schema/context"
xmlns:batch=" http://www.springframework.org/schema/batch"
xsi:schemaLocation="http://www.springframework.org/schema/context
http://www.springframework.org/schema/context/spring-context-4.2.xsd
http://www.springframework.org/schema/beans http://www.springframework.org/schema/beans/spring-beans-4.2.xsd">

<!--将配置文件读取到容器中,交给 Spring 管理 -->
<bean id="propertyConfigure"
class="org.springframework.beans.factory.config.PropertyPlaceholderConfigurer">
<property name="locations">
<list>
<!-- classpath 即 src 根目录 -->
<value>classpath:jdbc.properties</value>
```

```
</list>
</property>
</bean>

<! ——将 Druid 数据库连接池配置文件读取到容器 ——>
<bean id="druiddataSource"
class="com. alibaba. druid. pool. DruidDataSource" destroy-method="close">
<property name="driverClassName" value=" ${driver}" />
<property name="url" value=" ${url}" />
<property name="username" value=" ${username}" />
<property name="password" value=" ${password}" />
<property name="initialSize" value="1" />
<property name="minIdle" value="1" />
<property name="maxActive" value="10" />
<property name="maxWait" value="10000" />
<property name="timeBetweenEvictionRunsMillis" value="60000" />
<property name="minEvictableIdleTimeMillis" value="300000" />
<property name="testWhileIdle" value="true" />
</bean>

<! ——自动扫描和注册 bean ——>
<context:component-scan base-package="cn. hbmy. p13">
<! —这里暂不注解为 controller 的类型,controller 放到 Spring MVC 扫描 ——>
<context:exclude-filter type="annotation"
expression="org. springframework. stereotype. Controller" />
</context:component-scan>

<! ——配置 Session 工厂 ——>
< bean id = " sqlSessionFactory " class = " org. mybatis. spring.
SqlSessionFactoryBean">
<! —— 引用上面已经配置好的数据库连接池 ——>
<property name="dataSource" ref="druiddataSource" />
<! —— Mybatis 的配置文件路径 ——>
<property name="configLocation" value="classpath:Mybatis-Config. xml" />
<! —— 配置 mybatis 分页插件 PageHelper ——>
```

```
<property name="plugins">
<array>
<bean class="com. github. pagehelper. PageInterceptor">
<property name="properties">
<! -- 什么都不配,使用默认的配置 -->
<value></value>
</property>
</bean>
</array>
</property>
</bean>

<! --将 Mapper 接口生成代理注入到 Spring -->
<bean class="org. mybatis. spring. mapper. MapperScannerConfigurer">
<property name="basePackage" value="cn. hbmy. p13. mapper" />
<property name="sqlSessionFactoryBeanName" value="sqlSessionFactory"/>
</bean>

</beans>
```

4. Mybatis-Config. xml.

```
<? xml version="1. 0" encoding="UTF-8"? >
<! DOCTYPE configuration PUBLIC "-//mybatis. org//DTD Config 3. 0//EN"
  "http://mybatis. org/dtd/mybatis-3-config. dtd">
<configuration>
<typeAliases>
    <typeAlias type="cn. hbmy. p13. entity. User" alias="User" />
    <typeAlias type="cn. hbmy. p13. entity. Student" alias="Student" />
  </typeAliases>
  <mappers>
    <mapper resource="cn/hbmy/p13/mapper/UserMapper. xml" />
    <mapper resource="cn/hbmy/p13/mapper/StudentMapper. xml" />
  </mappers>
</configuration>
```

5. log4j2. xml.

```
<? xml version="1.0" encoding="UTF-8"? >
<configuration status=" debug ">
<Properties>
<Property name="backupFilePath">d:/logs/</Property>
</Properties>

<! --定义所有的 appender -->
<appenders>
<! --输出到控制台的配置 -->
<Console name="Console" target="SYSTEM_OUT">
<ThresholdFilter level="debug" onMatch="ACCEPT" onMismatch="DENY"
/>
<! --输出日志的格式 -->
<PatternLayout pattern="%d{HH:mm:ss.SSS} %-5level
%class{36} %L %M - %msg%xEx%n" />
</Console>

<! --输出到滚动保存文件的配置-->
<RollingFile name="RollingFileInfo" fileName="${backupFilePath}/info.log"
filePattern="${backupFilePath}/$${date:yyyy-MM}/info-%d{yyyy-MM
-dd}-%i.log">
<Filters>
<ThresholdFilter level="INFO" />
<ThresholdFilter level="WARN" onMatch="DENY"
onMismatch="NEUTRAL" />
</Filters>
<PatternLayout
pattern="[%d{HH:mm:ss:SSS}] [%p] - %l - %m%n" />
<Policies>
<TimeBasedTriggeringPolicy />
<SizeBasedTriggeringPolicy size="100 MB" />
</Policies>
<DefaultRolloverStrategy max="20" min="0" />
</RollingFile>
```

```
</appenders>

<!--定义 logger -->
<loggers>
<Logger name="cn.hbmy.p13.controller" level="INFO" additivity="false">
</Logger>
<Root level="all">
<appender-ref ref="Console" />
<appender-ref ref="RollingFileInfo" />
</Root>
</loggers>
</configuration>
```

6. dispatcher-servlet.xml.

```
<?xml version="1.0" encoding="UTF-8"?>
<beans xmlns="http://www.springframework.org/schema/beans"
xmlns:xsi="http://www.w3.org/2001/XMLSchema-instance"
xmlns:aop="http://www.springframework.org/schema/aop"
xmlns:context="http://www.springframework.org/schema/context"
xmlns:mvc="http://www.springframework.org/schema/mvc"
xsi:schemaLocation="http://www.springframework.org/schema/beans http://www.springframework.org/schema/beans/spring-beans.xsd
http://www.springframework.org/schema/context http://www.springframework.org/schema/context/spring-context-4.3.xsd
http://www.springframework.org/schema/aop http://www.springframework.org/schema/aop/spring-aop-4.3.xsd
http://www.springframework.org/schema/mvc http://www.springframework.org/schema/mvc/spring-mvc-4.3.xsd">

<!--开启注解 -->
<mvc:annotation-driven />

<!--启动自动扫描 -->
<context:component-scan
base-package="cn.hbmy.p13">
```

```
<! —— 制定扫包规则 ,只扫描使用@Controller 注解的 JAVA 类 ——>
<context:include—filter type="annotation"
expression="org. springframework. stereotype. Controller" />
</context:component—scan>

<! ——增加 AOP 的支持,并注册 LogAspect ——>
<aop:aspectj—autoproxy proxy—target—class="true" />
<bean class="cn. hbmy. p13. interceptor. LogAspect" />

<! ——访问静态资源 ——>
<mvc:resources location="/resources/css/" mapping="/css_mapping/ * * " />
<mvc:resources location="/resources/js/" mapping="/js_mapping/ * * " />
<mvc:resources location="/resources/img/"mapping="/img_mapping/ * * " />
<! ——配置视图解析器 ——>
<bean class="org. springframework. web. servlet. view. UrlBasedViewResolver">
<property name="viewClass"
value="org. springframework. web. servlet. view. JstlView"></property>
<property name="prefix" value="/WEB—INF/jsp/"></property>
<property name="suffix" value=". jsp"></property>
</bean>

<! —设置拦截器 ——>
<mvc:interceptors>
<! —若有多个拦截器,顺序执行 ——>
<mvc:interceptor>
<! —— / * * 的意思是所有文件夹及里面的子文件夹 ——>
<mvc:mapping path="/ * * / * " />
<mvc:exclude—mapping path="/ * * / * . css" />
<mvc:exclude—mapping path="/ * * / * . js" />
<mvc:exclude—mapping path="/ * * / * . jpg" />
<mvc:exclude—mapping path="/ * * / * . gif" />
<bean class="cn. hbmy. p13. interceptor. LoginInterceptor"></bean>
</mvc:interceptor>
</mvc:interceptors>
</beans>
```

7. decoractors. xml.

```
<? xml version="1.0" encoding="UTF-8"? >
<! --装饰文件默认存放目录 -->
<decorators defaultdir="/WEB-INF/decorators">
    <excludes>
        <pattern>/login.do</pattern>
        <! -- <pattern>/exclude/ *</pattern> -->
    </excludes>
<! --指定了一个名为 main 的装饰器  -->
    <decorator name="main" page="main.jsp">
        <pattern>/student/ *</pattern>
        <pattern>/user/ *</pattern>
        <pattern>/common/welcome.do</pattern>
    </decorator>
</decorators>
```

8. StudentMapper. xml.

```
<? xml version="1.0" encoding="UTF-8"? >
<! DOCTYPE mapper PUBLIC "-//mybatis.org//DTD Mapper 3.0//EN" "
http://mybatis.org/dtd/mybatis-3-mapper.dtd">
<mapper namespace="cn.hbmy.p13.mapper.StudentMapper">
<resultMap id="StudentMap" type="cn.hbmy.p13.entity.Student">
<id column="stuNum" jdbcType="VARCHAR" property="stunum" />
<result column="stuName" jdbcType="VARCHAR" property="stuname" />
<result column="stuSex" jdbcType="VARCHAR" property="stusex" />
<result column="stuAge" jdbcType="INTEGER" property="stuage" />
<result column="stuMajor" jdbcType="VARCHAR" property="stumajor" />
</resultMap>
<sql id="Base_Column_List">
stuNum, stuName, stuSex, stuAge, stuMajor
</sql>
<select id="selectStusByConditionListPage"
parameterType="String" resultMap="StudentMap">
 ${sql}
```

```
</select>

<select id="selectByPrimaryKey" parameterType="java. lang. String"
resultMap="StudentMap">
select
<include refid="Base_Column_List" />
from edu_stu
where stuNum = #{stunum,jdbcType=VARCHAR}
</select>

<delete id="delete" parameterType="String">
delete from edu_stu
where
stuNum = #{stunum,jdbcType=VARCHAR}
</delete>

<insert id="insert" parameterType="Student">
insert into edu_stu (stuNum,
stuName, stuSex,
stuAge, stuMajor)
values (#{stunum,jdbcType=VARCHAR},
#{stuname,jdbcType=VARCHAR},
#{stusex,jdbcType=VARCHAR},
#{stuage,jdbcType=INTEGER}, #{stumajor,jdbcType=VARCHAR})
</insert>

<update id="update" parameterType="Student">
update edu_stu
set stuName =
#{stuname,jdbcType=VARCHAR},
stuSex = #{stusex,jdbcType=VARCHAR},
stuAge = #{stuage,jdbcType=INTEGER},
stuMajor =
#{stumajor,jdbcType=VARCHAR}
where stuNum = #{stunum,jdbcType=VARCHAR}
```

```
</update>
</mapper>
```

9. UserMapper. xml.

```
<? xml version="1. 0" encoding="UTF-8"? >
<! DOCTYPE mapper PUBLIC "-//mybatis. org//DTD Mapper 3. 0//EN" "
http://mybatis. org/dtd/mybatis-3-mapper. dtd">
<mapper namespace="cn. hbmy. p13. mapper. UserMapper">
<resultMap id="UserMap" type="cn. hbmy. p13. entity. User">
<id column="userId" jdbcType="INTEGER" property="userid" />
<result column="userName" jdbcType="VARCHAR"property="username" />
<result column="userPwd" jdbcType="VARCHAR" property="userpwd" />
<result column="userType" jdbcType="VARCHAR" property="usertype" />
</resultMap>

<select id="selectByUsername" parameterType="java. lang. String" resultMap
="UserMap">
select * from edu _ user where userName = #{username, jdbcType =
VARCHAR}
</select>

<insert id="insert" parameterType="User">
insert into edu_user(userName, userPwd,userType)
values (#{username,jdbcType=VARCHAR},
#{userpwd,jdbcType=VARCHAR},#{usertype,jdbcType=VARCHAR})
</insert>

<update id="updateByPrimaryKey" parameterType="User">
update edu_user
set userName = #{username,jdbcType=VARCHAR},
userPwd = #{userpwd,jdbcType=VARCHAR},
userType = #{usertype,jdbcType=VARCHAR}
where userId = #{userid,jdbcType=INTEGER}
</update>
</mapper>
```

二、程序代码

1. 实体类(包 cn. hbmy. p13. entity).

(1)Student. java。

```
package cn.hbmy.p13.entity;

public class Student {
      private String stunum;
      private String stuname;
      private String stusex;
      private Integer stuage;
      private String stumajor;
   //省略 setter 和 getter 方法
……
}
```

(2)User. java.

```
package cn.hbmy.p13.entity;

public class User {
      private Integer userid;
      private String username;
      private String userpwd;
private String usertype;
   //省略 setter 和 getter 方法
……
}
```

2. DAO 接口(包 cn. hbmy. p13. mapper).

(1)StudentMapper. java.

```
package cn.hbmy.p13.mapper;

import java.util.List;
import org.apache.ibatis.annotations.Param;
import cn.hbmy.p13.entity.Student;
```

```
public interface StudentMapper {
List<Student> selectStusByConditionListPage(@Param(value="sql") String sql);
    int delete(String stuNum);
    int insert(Student stu);
    Student selectByPrimaryKey(String stunum);
    int update(Student stu);
}
```

(2)UserMapper. java.

```
package cn.hbmy.p13.mapper;

import cn.hbmy.p13.entity.User;

public interface UserMapper {
    int insert(User user);
    User selectByUsername(String username);
    int updateByPrimaryKey(User user);
}
```

3. Service 接口和接口实现类(包 cn. hbmy. p13. service)。

(1)IStudentService. java.

```
package cn.hbmy.p13.service;

import com.github.pagehelper.PageInfo;
import cn.hbmy.p13.entity.Student;

public interface IStudentService {
public PageInfo<Student> getAllStuByCondition(Integer pageNo, Integer pageSize,String condition);
public Student getStuByStunum(String stunum);
public void addStu(Student student);
public void updateStu(Student student);
public void deleteStu(String stunum);
```

```
public Student loadStu(String userNum);
}
```

(2)IUserService. java.

```
package cn.hbmy.p13.service;

import cn.hbmy.p13.entity.User;

public interface IUserService {
public User checkUser(User user);
public boolean registUser(User user);
public void updateUser(User user);
public boolean login(String username,String password);
}
```

(3)StudentService. java.

```
package cn.hbmy.p13.service;

import java.util.List;
import org.springframework.beans.factory.annotation.Autowired;
import org.springframework.stereotype.Service;
import com.github.pagehelper.PageHelper;
import com.github.pagehelper.PageInfo;
import cn.hbmy.p13.entity.Student;
import cn.hbmy.p13.mapper.StudentMapper;

@Service
public class StudentService implements IStudentService {

@Autowired
private StudentMapper studentMapper;

public void addStu(Student student) {
Student stu = studentMapper.selectByPrimaryKey(student.getStunum());
if (stu != null)
```

```
System.out.println("要添加的学号已经存在");
else
studentMapper.insert(student);
}

public void updateStu(Student student) {
studentMapper.update(student);
}

public void deleteStu(String stunum) {
studentMapper.delete(stunum);


}

public Student loadStu(String stunum) {
return studentMapper.selectByPrimaryKey(stunum);
}

public PageInfo<Student> getAllStuByCondition(Integer pageNo, Integer
pageSize, String condition) {
String sql = "select * from edu_stu where " + condition;
pageNo = pageNo == null ? 1 : pageNo;
pageSize = pageSize == null ? 10 : pageSize;
PageHelper.startPage(pageNo, pageSize);
List<Student> stus = studentMapper.selectStusByConditionListPage(sql);
PageInfo<Student> page = new PageInfo<Student>(stus);
return page;
}

public Student getStuByStunum(String stunum) {
Student stu = studentMapper.selectByPrimaryKey(stunum);
return stu;
}
}
```

(4)UserService.java.

```
package cn.hbmy.p13.service;

import org.springframework.beans.factory.annotation.Autowired;
import org.springframework.stereotype.Service;
import cn.hbmy.p13.entity.User;
import cn.hbmy.p13.mapper.UserMapper;

@Service("userService")
public class UserService implements IUserService {

@Autowired
private UserMapper userMapper;

public User checkUser(User user) {
User u = userMapper.selectByUsername(user.getUsername());
if (u != null) {
String p1 = user.getUserpwd();
String p2 = u.getUserpwd();
String t1 = user.getUsertype();
String t2 = u.getUsertype();
if (p1.equals(p2) && t1.equals(t2))
return u;
else
return null;

} else
return null;
}

public boolean registUser(User user) {
User u = userMapper.selectByUsername(user.getUsername());
if (u == null) {
userMapper.insert(user);
return true;
```

```
} else
return false;
}

public void updateUser(User user) {
userMapper.updateByPrimaryKey(user);
}

public boolean login(String username, String password) {
User u = userMapper.selectByUsername(username);
boolean b = false;
if (u == null) {
System.out.println("登录用户不存在");
} else if (u.getUserpwd().equals(password))
b = true;
else
System.out.println("用户密码不正确");
return b;
}
}
```

4. Controller 控制器((包 cn. hbmy. p13. controller).

(1)IndexController. java.

```
package cn.hbmy.p13.controller;

import org.springframework.beans.factory.annotation.Autowired;
import org.springframework.stereotype.Controller;
import org.springframework.web.bind.annotation.RequestMapping;
import org.springframework.web.bind.annotation.RequestMethod;
import org.springframework.web.bind.annotation.RequestParam;
import org.springframework.web.servlet.ModelAndView;
import cn.hbmy.p13.entity.Student;
import cn.hbmy.p13.service.IStudentService;

@Controller
```

```
public class IndexController {

    @Autowired
    private IStudentService studentService;

    @RequestMapping(value = "/showloginjsp.do", method =
RequestMethod.GET)
    public String login( ) {
    return "common/login";
    }

    @RequestMapping(value = "/showregisterjsp.do", method =
RequestMethod.GET)
    public String register( ) {
    return "user/register";
    }

    @RequestMapping(value = "/common/welcome.do", method =
{ RequestMethod.POST, RequestMethod.GET })
    public String welcome( ) {
    return "common/welcome";
    }

    @RequestMapping(value = "student/showQueryJsp.do", method =
RequestMethod.GET)
    public String showQueryJsp( ) throws Exception {
    return "/student/query";
    }

    @RequestMapping(value = "student/showDeleteJsp.do", method =
RequestMethod.GET)
    public String showDeleteJsp( ) throws Exception {
    return "/student/delete";
    }
```

```
@RequestMapping(value = "student/showCreateJsp.do", method =
RequestMethod.GET)
public String showCreateJsp( ) throws Exception {
return "/student/create";
}

@RequestMapping(value = "student/showUpdatelistJsp.do", method =
RequestMethod.GET)
public String showUpdatelistJsp( ) throws Exception {
return "/student/updatelist";
}

@RequestMapping(value="student/showUpdatepageJsp.do",method =
RequestMethod.GET)
public ModelAndView showUpdatepageJsp(@RequestParam("stunum") String
stunum) throws Exception {
Student stu = studentService.loadStu(stunum);
return new ModelAndView("/student/update", "stu", stu);
}

}
```

(2)StudentController.java.

```
package cn.hbmy.p13.controller;

import java.io.PrintWriter;
import javax.servlet.http.HttpServletResponse;
import javax.servlet.http.HttpSession;
import org.json.JSONObject;
import org.springframework.beans.factory.annotation.Autowired;
import org.springframework.stereotype.Controller;
import org.springframework.ui.Model;
import org.springframework.validation.BindingResult;
import org.springframework.validation.annotation.Validated;
import org.springframework.web.bind.annotation.PathVariable;
```

```
import org.springframework.web.bind.annotation.RequestMapping;
import org.springframework.web.bind.annotation.RequestMethod;
import org.springframework.web.bind.annotation.RequestParam;
import com.github.pagehelper.PageInfo;
import cn.hbmy.p13.entity.Student;
import cn.hbmy.p13.service.IStudentService;

@Controller
public class StudentController {

@Autowired
private IStudentService studentService;

@RequestMapping(value = "student/create.do", method = RequestMethod.POST)
public void create(@RequestParam("stunum") String stunum, @RequestParam("stuname") String stuname,@RequestParam("stusex") String stusex, @RequestParam("stuage") String stuage,
@RequestParam("stumajor") String stumajor, HttpServletResponse response) throws Exception {
JSONObject json = new JSONObject();
PrintWriter out = response.getWriter();
Student stu = new Student();
stu.setStunum(stunum);
stu.setStuname(stuname);
stu.setStusex(stusex);
stu.setStuage(Integer.valueOf(stuage));
stu.setStumajor(stumajor);
if (studentService.loadStu(stunum) != null) {
json.put("result", "exist");
out.print(json);
} else {
studentService.addStu(stu);
Student s = studentService.getStuByStunum(stu.getStunum());
if (s != null) {
```

```
json.put("result", "success");
} else {
json.put("result", "error");
}
out.print(json);
}
}

@RequestMapping(value = "student/update.do", method = RequestMethod.POST)
public void update(@RequestParam("stunum") String stunum, @RequestParam("stuname") String stuname, @RequestParam("stusex") String stusex, @RequestParam("stuage") String stuage, @RequestParam("stumajor") String stumajor, HttpServletResponse response) throws Exception {
JSONObject json = new JSONObject();
PrintWriter out = response.getWriter();
Student stu = new Student();
stu.setStunum(stunum);
stu.setStuname(stuname);
stu.setStusex(stusex);
stu.setStuage(Integer.valueOf(stuage));
stu.setStumajor(stumajor);
studentService.updateStu(stu);
json.put("result", "success");
out.print(json);
}

@RequestMapping(value = "student/queryByCondition.do", method = {RequestMethod.POST, RequestMethod.GET})
public String queryByCondition(@RequestParam(required = false, defaultValue = "1") int page, @RequestParam(required = false, defaultValue = "10") int pageSize, @RequestParam(required = false) String field, @RequestParam(required = false) String op, @RequestParam(required = false) String value, HttpSession session, Model model) throws Exception {
PageInfo<Student> stus = new PageInfo<Student>();
```

```
String condition = (String) session.getAttribute("condition_1");
if (condition == null || condition == "") {
if ("greater".equals(op))
op = ">";
if ("less".equals(op))
op = "<";
if ("equal".equals(op))
op = "=";
if ("stuAge".equals(field))
condition = field + op + value;
else
condition = field + op + "'" + value + "'";
session.setAttribute("condition_1", condition);
}
stus = studentService.getAllStuByCondition(page, pageSize, condition);
model.addAttribute("dataList", stus);
return "student/query";
}

@RequestMapping(value = "student/queryByCondition_u.do", method
={RequestMethod.POST, RequestMethod.GET })
public String queryByCondition_u(@RequestParam(required = false,
defaultValue = "1") int page, @RequestParam(required = false, defaultValue = "
10") int pageSize,  @RequestParam(required = false) String field, @RequestParam
(required = false) String op,  @RequestParam(required = false) String value,
 HttpSession session, Model model) throws Exception {
PageInfo<Student> stus = new PageInfo<Student>();
String condition = (String) session.getAttribute("condition_2");
if (condition == null || condition == "") {
if ("greater".equals(op))
op = ">";
if ("less".equals(op))
op = "<";
if ("equal".equals(op))
op = "=";
```

```
if ("stuAge". equals(field))
condition = field + op + value;
else
condition = field + op + "'" + value + "'";
session. setAttribute("condition_2", condition);
}
stus = studentService. getAllStuByCondition(page, pageSize, condition);
model. addAttribute("dataList", stus);
return "student/updatelist";
}

@RequestMapping(value = "student/queryByCondition_d. do", method = {RequestMethod. POST, RequestMethod. GET})
public String queryByCondition_d(@RequestParam(required = false, defaultValue = "1") int page, @RequestParam(required = false, defaultValue = "10") int pageSize, @RequestParam(required = false) String field, @RequestParam(required = false) String op, @RequestParam(required = false) String value, HttpSession session, Model model) throws Exception {
PageInfo<Student> stus = new PageInfo<Student>();
String condition = (String) session. getAttribute("condition_3");
if (condition == null || condition == "") {
if ("greater". equals(op))
op = ">";
if ("less". equals(op))
op = "<";
if ("equal". equals(op))
op = "=";
if ("stuAge". equals(field))
condition = field + op + value;
else
condition = field + op + "'" + value + "'";
session. setAttribute("condition_3", condition);
}
stus = studentService. getAllStuByCondition(page, pageSize, condition);
model. addAttribute("dataList", stus);
```

```
return "student/delete";
}

@RequestMapping(value = "/{id}/updatestudent", method = RequestMethod.
GET)
public String update(@PathVariable String id, Model model) {
model. addAttribute(studentService. loadStu(id));
return "student/update";
}

@RequestMapping(value = "/{id}/updatestudent", method = RequestMethod.
POST)
public String update (@ PathVariable String id, @ Validated Student student,
BindingResult br, Model model) {
if (br. hasErrors( )) {
return "article/update";
}
Student tu = studentService. loadStu(id);
tu. setStuname(student. getStuname( ));
tu. setStunum(student. getStunum( ));
tu. setStuage(student. getStuage( ));
tu. setStusex(student. getStusex( ));
tu. setStumajor(student. getStumajor( ));
studentService. updateStu(tu);
return "redirect:/student/articles";
}

@ RequestMapping ( value = " student/delete. do", method = RequestMethod.
GET)
public String delete(@RequestParam("stunum") String stunum) {
studentService. deleteStu(stunum);
return "redirect:/student/showDeleteJsp. do";
}
}
```

(3)UserController. java.

```
package cn. hbmy. p13. service;

import org. springframework. beans. factory. annotation. Autowired;
import org. springframework. stereotype. Service;

import cn. hbmy. p13. entity. User;
import cn. hbmy. p13. mapper. UserMapper;

@Service("userService")
public class UserService implements IUserService {

@Autowired
private UserMapper userMapper;

@Override
public User checkUser(User user) {
User u = userMapper. selectByUsername(user. getUsername( ));
if (u ! = null) {
String p1 = user. getUserpwd( );
String p2 = u. getUserpwd( );
String t1 = user. getUsertype( );
String t2 = u. getUsertype( );
if (p1. equals(p2) && t1. equals(t2))
return u;
else
return null;

} else
return null;
}

public boolean registUser(User user) {
User u = userMapper. selectByUsername(user. getUsername( ));
if (u == null) {
```

```
userMapper. insert(user);
return true;
} else
return false;
}

public void updateUser(User user) {
userMapper. updateByPrimaryKey(user);
}

public boolean login(String username, String password) {
User u = userMapper. selectByUsername(username);
boolean b = false;
if (u == null) {
System. out. println("登录用户不存在");
} else if (u. getUserpwd( ). equals(password))
b = true;
else
System. out. println("用户密码不正确");
return b;
}


}
```

三、前端页面

1. jsp/common 目录.

(1)login. jsp.

```
<%@ page contentType="text/html; charset=UTF-8"pageEncoding="UTF-8"%>
<html>
<head>
<script src="<%=request. getContextPath( )%>/js_mapping/jquery. js"></script>
<script type="text/javascript">
    $(function ( ){
```

```
        //登录
        $("#login").click(function ( ) {
    var username = document.getElementById('username').value;
    var userpwd = document.getElementById('userpwd').value;
    var usertype = document.getElementById('usertype').value;
    if(username==null || username=="" || userpwd==null ||
    userpwd=="" || usertype==null || usertype=="") {
      alert("用户名、密码和用户类型不能为空!");
      }else
              $.ajax({
                  type:"post",
                  url:"<%=request.getContextPath( )%>/user/login.do",
    dataType : "json",
    data : {
    "username" : $("#username").val( ),"userpwd" : $("#userpwd").val( ),
    "usertype" : $("#usertype").val( )},
    success : function(data) {
    if(data.result == "success") {
    window.location.href = 'common/welcome.do';} elsealert("用户名、密码或类型
错误!");},
    error : function( ) {
    alert("网络错误!");
    }
    });
    });
    });
    </script>
    <title>学生信息管理系统—登录</title>
    </head>

    <body>
    <div align="center" style="width: 100%; float: left; margin-top: 10px;">
    <div class="box">
    用户名:<input type="text" id="username" name="username" id="username" />
    密码:<input type="password" id="userpwd" name="userpwd" id="userpwd" />
```

```
用户类型：<select name="usertype" id="usertype">
<option value="T">老师</option>
<option value="S">学生</option>
</select> <input type="button" value="登录" id="login" /> <input type="button" value="注册" id="register"
onclick="window.location.href='<%=request.getContextPath()%>/showregisterjsp.do'" />
</div>
</div>
</body>
</html>
```

(2)resource.jsp.

```
<%@ page import="java.util.*" pageEncoding="UTF-8"%>
<!--公共资源CSS,JS -->
<link rel="stylesheet" type="text/css" href="<%=request.getContextPath()%>/css_mapping/left.css"/>
<link rel="stylesheet" type="text/css" href="<%=request.getContextPath()%>/css_mapping/table.css"/>
<script type="text/javascript" src="<%=request.getContextPath()%>/js_mapping/jquery.js">
</script>
<script type="text/javascript" src="<%=request.getContextPath()%>/js_mapping/left.js">
</script>
```

(3)welcome.jsp.

```
<%@ page contentType="text/html; charset=UTF-8" pageEncoding="UTF-8"%>
<html>
<head>
<title>学生信息管理系统</title>
</head>
<body>
<h1>欢迎访问学生信息管理系统！</h1>
```

```
</body>
</html>
```

2. jsp/student 目录.

(1)create. jsp.

```
<%@ page contentType="text/html; charset=UTF-8" pageEncoding="UTF-8"%>
<html>
<head>
<title>新增学生信息</title>
<script type="text/javascript">
$(function (){
var _age = document.getElementById("stuage");
for(var i=10; i<=40; i++){
      var _option = new Option(i,i);
      _age.options.add(_option);
   }

$("#create").click(function () {
var stuname = document.getElementById('stuname').value;
var stunum = document.getElementById('stunum').value;
var stusex = document.getElementById('stusex').value;
var stuage = document.getElementById('stuage').value;
var stumajor = document.getElementById('stumajor').value;
if(stunum==null || stunum=="" || stuname==null || stuname=="")
alert("学生的学号和姓名不能为空!");
else if(stunum.length! =9)
alert("学号长度9个字符!")
   else
                $.ajax({
                  type:"post",
                  url:"<%=request.getContextPath()%>/student/create.do",
dataType : "json",
data : {
"stunum" : $("#stunum").val(),
```

```
"stuname" : $("#stuname").val( ),
"stusex" : $("#stusex").val( ),
"stuage" : $("#stuage").val( ),
"stumajor" : $("#stumajor")
.val( )
},
success : function(data) {
if (data.result == "success")
alert("新增学生信息成功!");
else if (data.result == "exist")
alert("学号" + stunum
+ "已存在!");
else
alert("新增学生信息失败!");
},
error : function( ) {
alert("网络错误!");
}
});
});
});
</script>
</head>
<body>
<table>
<tr><td>学号:</td>
<td><input type="text" id="stunum" name="stunum" /></td></tr>
<tr><td>姓名:</td>
<td><input type="text" id="stuname" name="stuname" /></td></tr>
<tr><td>性别:</td>
<td><select id="stusex" name="stusex">
<option value="男">男</option>
<option value="女">女</option>
</select></td></tr>
<tr><td>年龄:</td>
```

```
<td><select id="stuage" name="stuage">
</select></td></tr>
<tr><td>专业:</td>
<td><select id="stumajor" name="stumajor">
<option value="计算机科学与技术">计算机科学与技术</option>
<option value="信息安全">信息安全</option>
<option value="数字媒体">数字媒体</option>
</select></td></tr>
<tr><td><input type="button" name="create" id="create" value="新增" /></
td>
<td><input type="button" name="back" value="返回"
onclick="javascript:window.history.back(-1);"></td></tr>
</table>
</body>
</html>
```

(2)delete.jsp.

```
<%@ page contentType="text/html; charset=UTF-8" pageEncoding="UTF
-8"%>
<%--引入 JSTL 标签库 --%>
<%@ taglib uri="http://java.sun.com/jsp/jstl/core" prefix="c"%>
<html>
<head>
<title>删除学生信息</title>
</head>
<body>
<form action="<%=request.getContextPath()%>/student/queryByCondition_d.do"
method="post">
<table style="width: 100%;">
<tr><td>查询需要删除的记录</td>
<td>输入条件查询:</td>
<td><select name="field">
<option value="stunum">学号</option>
<option value="stuname">姓名</option>
<option value="stusex">性别</option>
```

```
<option value="stuage">年龄</option>
<option value="stumajor">专业</option>
</select>
<select name="op">
<option value="greater">></option>
<option value="less">&lt;</option>
<option value="equal">=</option>
</select>
<input type="text" name="value" /></td>
<td><input type="submit" value="查看记录" /></td></tr>
</table>
</form>
<c:if test="${dataList ne null}">
<table id="students">
<tr><th>学号</th><th>姓名</th><th>性别</th>
<th>年龄</th><th>专业</th><th>操作</th></tr>
<c:forEach var="stu" items="${dataList.list}">
<tr class='alt'>
<td>${stu.stunum}</td><td>${stu.stuname}</td>
<td>${stu.stusex}</td><td>${stu.stuage}</td>
<td>${stu.stumajor}</td>
<td>
<a href="javascript:if(confirm('确实要删除该内容吗?')) location='<%=request.getContextPath()%>/student/delete.do?stunum=${stu.stunum}'">删除</a></td>
</tr>
</c:forEach>
</table>
共${dataList.total}条记录,共${dataList.pages}页
<c:if test="${dataList.pageNum ne dataList.firstPage}">
<ahref="<%=request.getContextPath()%>/student/queryByCondition_d.do?page=${dataList.firstPage}">首页</a>
</c:if>
<c:if test="${dataList.pageNum eq dataList.firstPage}">首页</c:if>
<c:if test="${dataList.pageNum gt dataList.firstPage}">
```

```
<ahref="<%=request.getContextPath()%>/student/queryByCondition_d.do? page=${dataList.prePage}">≪上一页</a>
</c:if>
<c:if test="${dataList.pageNum eq dataList.firstPage}">≪上一页</c:if>
<c:if test="${dataList.pageNum lt dataList.lastPage}">
<a href="<%=request.getContextPath()%>/student/queryByCondition_d.do? page=${dataList.nextPage}">下一页≫</a>
</c:if>
<c:if test="${dataList.pageNum eq dataList.lastPage}">下一页≫</c:if>
<c:if test="${dataList.pageNum ne dataList.lastPage}">
<a href="<%=request.getContextPath()%>/student/queryByCondition_d.do? page=${dataList.lastPage}">最后页</a>
</c:if>
<c:if test="${dataList.pageNum eq dataList.lastPage}"> 最后页</c:if>
</c:if>
</body>
</html>
```

(3)query.jsp.

```
<%@ page contentType="text/html; charset=UTF-8" pageEncoding="UTF-8"%>
<%@ taglib uri="http://java.sun.com/jsp/jstl/core" prefix="c"%>
<html>
<head>
<title>浏览学生信息</title>
</head>
<body>
<form action="<%=request.getContextPath()%>/student/queryByCondition.do" method="post">
<table style="width: 100%;">
<tr><td>输入条件查询:</td>
<td><select name="field">
<option value="stunum">学号</option>
<option value="stuname">姓名</option>
<option value="stusex">性别</option>
```

```
<option value="stuage">年龄</option>
<option value="stumajor">专业</option></select>
<select name="op">
<option value="greater">></option>
<option value="less">&lt;</option>
<option value="equal">=</option></select>
<input type="text" name="value" /></td>
<td><input type="submit" value="查看记录" /></td></tr>
</table>
</form>
<c:if test="${dataList ne null}">
<table id="students">
<tr><th>学号</th><th>姓名</th><th>性别</th>
<th>年龄</th><th>专业</th></tr>
<c:forEach var="stu" items="${dataList.list}">
<tr class='alt'>
<td>${stu.stunum}</td><td>${stu.stuname}</td>
<td>${stu.stusex}</td><td>${stu.stuage}</td>
<td>${stu.stumajor}</td>
</tr>
</c:forEach>
</table>
共${dataList.total}条记录,共${dataList.pages}页
<c:if test="${dataList.pageNum ne dataList.firstPage}">
<a href="<%=request.getContextPath()%>/student/queryByCondition.do?page= ${dataList.firstPage}">首页</a></c:if>
<c:if test="${dataList.pageNum eq dataList.firstPage}">首页</c:if>
<c:if test="${dataList.pageNum gt dataList.firstPage}">
<a href="<%=request.getContextPath()%>/student/queryByCondition.do?page= ${dataList.prePage}">≪上一页</a></c:if>
<c:if test="${dataList.pageNum eq dataList.firstPage}">≪上一页</c:if>
<c:if test="${dataList.pageNum lt dataList.lastPage}">
<a href="<%=request.getContextPath()%>/student/queryByCondition.do?page= ${dataList.nextPage}">下一页≫</a></c:if>
<c:if test="${dataList.pageNum eq dataList.lastPage}">下一页≫</c:if>
```

```
<c:if test="${dataList.pageNum ne dataList.lastPage}">
<a href="<%=request.getContextPath()%>/student/queryByCondition.do?page = ${dataList.lastPage}">最后页</a></c:if>
<c:if test="${dataList.pageNum eq dataList.lastPage}"> 最后页</c:if>
</c:if>
</body>
</html>
```

(4)update.jsp.

```
<%@ page contentType="text/html; charset=UTF-8" pageEncoding="UTF-8"%>
<html>
<head>
<title>修改学生信息</title>
<script type="text/javascript">
    $(function(){
        var sex="${stu.stusex}";
        var major="${stu.stumajor}";
        var age="${stu.stuage}";
var _age = document.getElementById("stuage");
for(var i=10; i<=40; i++){
    var _option = new Option(i,i);
    _age.options.add(_option);
  }
 $("#stuage option[value='"+age+"']").attr("selected","selected");
 $("#stusex option[value='"+sex+"']").attr("selected","selected");
 $("#stumajor option[value='"+major+"']").attr("selected","selected");

 $("#update").click(function() {
var stuname = document.getElementById('stuname').value;
var stusex = document.getElementById('stusex').value;
var stuage = document.getElementById('stuage').value;
var stumajor = document.getElementById('stumajor').value;
if(stuname==null || stuname=="")
alert("学生的姓名不能为空!");
```

```
else
              $.ajax({
                type:"post",
                url:"<%=request.getContextPath()%>/student/update.do",
dataType : "json",
data : {
"stunum" : $("#stunum").val(),
"stuname" : $("#stuname").val(),
"stusex" : $("#stusex").val(),
"stuage" : $("#stuage").val(),
"stumajor" : $("#stumajor").val()
},
success : function(data) {
if (data.result == "success")
alert("修改学生信息成功!");
else
alert("修改学生信息失败!");
},
error : function() {
alert("网络错误!");
}
});
});
});
</script>
</head>
<body>
<table>
<tr><td>学号:</td>
<td><input type="text" id="stunum" name="stunum"
disabled="disabled" value="${stu.stunum}" /></td></tr>
<tr><td>姓名:</td>
<td><input type="text" id="stuname" name="stuname"
value="${stu.stuname}" /></td></tr>
<tr><td>性别:</td>
```

```
<td><select id="stusex" name="stusex">
<option value="男">男</option>
<option value="女">女</option></select></td></tr>
<tr> <td>年龄:</td> <td><select id="stuage" name="stuage"> </
select></td></tr>
<tr> <td>专业:</td> <td>
<select id="stumajor" name="stumajor">
<option value="计算机科学与技术">计算机科学与技术</option>
<option value="信息安全">信息安全</option>
<option value="数字媒体">数字媒体</option></select></td></tr>
<tr><td><input type="button" name="create" id="update" value="修改"
/></td>
<td><input type="button" name="back" value="返回"
onclick="javascript:window.history.back(-1);"></td></tr>
</table>
</body>
</html>
```

(5)updatelist.jsp.

```
<%@ page contentType="text/html; charset=UTF-8" pageEncoding="UTF
-8"%>
<%@ taglib uri="http://java.sun.com/jsp/jstl/core" prefix="c"%>
<html>
<head>
<title>修改学生信息</title>
</head>
<body>
<form action="<%=request.getContextPath( )%>/student/queryByCondition_u.do"
method="post">
<table style="width: 100%;">
<tr><td>查询需要修改的记录</td>
<td>输入条件查询:</td>
<td><select name="field">
<option value="stunum">学号</option>
<option value="stuname">姓名</option>
```

```
<option value="stusex">性别</option>
<option value="stuage">年龄</option>
<option value="stumajor">专业</option>
</select> <select name="op">
<option value="greater">></option>
<option value="less">&lt;</option>
<option value="equal">=</option>
</select> <input type="text" name="value" /></td>
<td><input type="submit" value="查看记录" /></td></tr>
</table>
</form>
<c:if test="${dataList ne null}">
<table id="students">
<tr><th>学号</th><th>姓名</th><th>性别</th>
<th>年龄</th><th>专业</th><th>操作</th></tr>
<c:forEach var="stu" items="${dataList.list}">
<tr class='alt'>
<td>${stu.stunum}</td><td>${stu.stuname}</td>
<td>${stu.stusex}</td><td>${stu.stuage}</td>
<td>${stu.stumajor}</td>
<td>
<a href="<%=request.getContextPath()%>/student/showUpdatepageJsp.do?stunum=${stu.stunum}">修改</a></td></tr>
</c:forEach>
</table>
共${dataList.total}条记录,共${dataList.pages}页
<c:if test="${dataList.pageNum ne dataList.firstPage}">
<a href="<%=request.getContextPath()%>/student/queryByCondition_u.do?page=${dataList.firstPage}">首页</a> </c:if>
<c:if test="${dataList.pageNum eq dataList.firstPage}">首页</c:if>
<c:if test="${dataList.pageNum gt dataList.firstPage}">
<a href="<%=request.getContextPath()%>/student/queryByCondition_u.do?page=${dataList.prePage}">≪上一页</a> </c:if>
<c:if test="${dataList.pageNum eq dataList.firstPage}">≪上一页</c:if>
<c:if test="${dataList.pageNum lt dataList.lastPage}">
```

```
<a href="<%=request. getContextPath( )%>/student/queryByCondition_u.do? page= ${dataList. nextPage}">下一页≫</a> </c:if>
<c:if test="${dataList. pageNum eq dataList. lastPage}">下一页≫</c:if>
<c:if test="${dataList. pageNum ne dataList. lastPage}">
<a href="<%=request. getContextPath( )%>/student/queryByCondition_u.do? page= ${dataList. lastPage}">最后页</a> </c:if>
<c:if test="${dataList. pageNum eq dataList. lastPage}"> 最后页</c:if>
</c:if>
</body>
</html>
```

3. jsp/user 目录。

(1)register. jsp.

```
<%@ page contentType="text/html; charset=UTF-8"pageEncoding="UTF-8"%>
<html>
<head>
<script src="<%=request. getContextPath( )%>/js_mapping/jquery. js"></script>
<title>用户注册</title>
<script type="text/javascript">
    $(function ( ){
        $("#register"). click(function ( ) {
var username = document. getElementById('username'). value;
var userpwd = document. getElementById('userpwd'). value;
var usertype = document. getElementById('usertype'). value;
if(username==null || username=="")
alert("用户名不能为空!");
else if(userpwd. length<6 || userpwd. length>10)
alert("密码长度 6～10 个字符!")
  else
            $. ajax({
                type:"post",
                url:"<%=request. getContextPath( )%>/user/register. do",
dataType : "json",
```

```
data : {
"username" : $("#username").val( ),
"userpwd" : $("#userpwd").val( ),
"usertype" : $("#usertype").val( ),
},
success : function(data) {
if (data.result == "success")
alert("新用户注册成功!");
else if (data.result == "error")
alert("用户名:" + username
+ "已存在!");
else
alert("新用户注册失败!");
},
error : function( ) {
alert("网络错误!");
}
});
});
});
</script>
</head>
<body>
<table>
<tr> <td>用户名:</td>
<td><input type="text" id="username" name="usernmae" /></td> </
tr>
<tr><td>密码:</td>
<td><input type="text" id="userpwd" name="userpwd" /></td></tr>
<tr><td>类型:</td>
<td><select id="usertype" name="usertype">
<option value="T">老师</option>
<option value="S">学生</option>
</select></td></tr>
<tr><td><input type="button" name="register" id="register" value="注
```

```
册" /></td>
    <td><input type="button" name="back" value="返回"
    onclick="javascript:window.history.back(-1);"></td> </tr>
    </table>
    </body>
    </html>
```

(2)updatepwd.jsp。

```
    <%@ page contentType="text/html; charset=UTF-8"pageEncoding="UTF
-8"%>
    <html>
    <head>
    <title>修改密码</title>
    <script type="text/javascript">
        $(function (){
            $("#update").click(function () {
    var pwd1 = document.getElementById('pwd1').value;
    var pwd2 = document.getElementById('pwd2').value;
    if(pwd1==null || pwd1==""){
      alert("旧密码不能为空!");}
    else if(pwd2==null || pwd2=="")
    alert("新密码不能为空!");
      else
                $.ajax({
                    type:"post",
                    url:"<%=request.getContextPath() %>/user/updatePwd.do",
                    dataType:"json",
    data:{"pwd1":$("#pwd1").val(),"pwd2":$("#pwd2").val()},
                    success:function(data) {
                    if(data.result=="success"){
                    alert("密码修改成功!");
                       }else
                         alert("旧密码错误,密码修改失败!");
                    },
                    error:function() {
```

```
                    alert("网络错误!");
                }
            });
        });
    });
    </script>
</head>

<body>
<table style="align-content: center; width: 700px;">
<tr><td>输入旧密码:</td>
<td><input type="text" id="pwd1" name="pwd1" /></td></tr>
<tr><td>输入新密码:</td>
<td><input type="text" id="pwd2" name="pwd2" /></td></tr>
<tr><td colspan="2"><input type="button" id="update" value="确定" />
</td>
</tr>
</table>
</body>
</html>
```

4. decorators 目录。

(1)footer. jsp.

```
<%@ page contentType="text/html; charset=UTF-8" pageEncoding="UTF-8"%>
<html>
<head></head>
<body>
<div align="center"
style="width: 100%; border-top: 1px solid; float: left; margin-top: 10px;">
CopyRight@2012-2018<br /> 学生信息管理系统
</div>
</body>
```

(2)header. jsp.

```
<%@ page contentType="text/html; charset=UTF-8"pageEncoding="UTF-8"%>
<html>
<body>
当前用户:${loginUser. username}
<a href="<%=request. getContextPath( )%>/logout. do">注销</a>
<h1>学生信息管理系统</h1>
</body>
</html>
```

(3)menu. jsp.

```
<%@ page pageEncoding="UTF-8"%>
<html>
<head></head>
<body>
<div class='box' style="display: inline; float: left;">
<div class="other_bar">
<ul class="menu"><li><h1 class="Left_Bar_title">
<i class="fa fa-truck"></i> <span>学生信息管理</span></h1>
<ul><li><p>
<a href="<%=request. getContextPath( )%>/student/showQueryJsp. do">查询学生信息</a>
</p></li> <li><p>
<a href="<%=request. getContextPath( )%>/student/showCreateJsp. do">新增学生信息</a>
</p></li><li><p>
<a href="<%=request. getContextPath( )%>/student/showUpdatelistJsp. do">修改学生信息</a>
</p></li><li><p>
<a href="<%=request. getContextPath( )%>/student/showDeleteJsp. do">删除学生信息</a>
</p></li></ul></li>
<li><h1 class="Left_Bar_title">
<i class="fa fa-user"></i> <span>个人信息</span></h1>
<ul><li><p>
```

```
<a href="<%=request.getContextPath( )%>/user/showUpdatePwdJsp.do"
>修改密码</a>
</p></li></ul></li></ul></div></div>
</body>
</html>
```

(4)main.jsp.

```
<%@ page contentType="text/html; charset=UTF-8"pageEncoding="UTF
-8"%>
<%@ include file="/WEB-INF/jsp/common/resource.jsp"%>
<%@ taglib prefix = "sitemesh" uri = "http://www.opensymphony.com/
sitemesh/decorator"%>
<html>
<head>
<style type="text/css">
html, body {
width: 98%;
height: 98%;
margin: 0;}
#Container {
position: relative;
width: 100%;
height: 100%;}
#Header {
position: relative;
width: 100%;
height: 10%;
background-color: #fff;}
#Content {
position: relative;
width: 100%;
height: 80%;}
#Content-left {
width: 20%;
height: 100%;
```

```
float: left;
background-color: #fff;}
#Content-right {
width: 79%;
height: 100%;
float: left;
background-color: #fff;}
#Footer {
position: relative;
width: 100%;
height: 10%;
background-color: #fff;}
</style>
<sitemesh:title />
<sitemesh:head />
</head>
<body>
<div id="Container">
<div id="Header"><%@include file="/WEB-INF/decorators/header.jsp"%
> </div>
<div id="Content"> <div id="Content-left">
<%@include file="/WEB-INF/decorators/menu.jsp"%></div>
<div id="Content-right"><sitemesh:body /> </div></div>
<div id="Footer"><%@include file="/WEB-INF/decorators/footer.jsp"%
></div>
</div>
</body>
</html>
```

四、静态资源

1. resources/css 目录。

(1)left. css.

```
* {
padding: 0;
margin: 0;
```

```
}
body {
font—size: 15px;
}
ul {
list—style—type: none;
}
h1 {
width: 100%;
font—size: 100%;
}
. box {
width: 100%;
border—bottom: 1px solid #fff;
background—color: #A7C942;
}
. menu>li {
border: 1px solid #ccc;
border—bottom: none;
color: #fff;
text—indent: 10px;
line—height: 30px;
font—size: 100%;
background—position: 170px 7px;
}
. menu>li. click {
background—position: 170px —27px;
}
. menu>li. hover {
background—color: #95c249;
}
. box p {
border—left: 6px solid #fff;
border—right: 1px solid #ccc;
line—height: 30px;
```

```
text—indent：30px；
color：＃000；
background—color：＃fff；
display：none；
transition：border 0. 2s linear 0s；
}
. box p. hover {
border—left：6px solid ＃A7C942；
}
. box a {
text—decoration：none；
color：＃000000；
}
```

(2)table. css.

```
＃students {
font—family："Trebuchet MS"，Arial，Helvetica，sans—serif；
width：100％；
border—collapse：collapse；
}
＃students td，＃students th {
font—size：1em；
border：1px solid ＃98bf21；
padding：3px 7px 2px 7px；
}
＃students th {
font—size：1. 1em；
text—align：left；
padding—top：5px；
padding—bottom：4px；
background—color：＃A7C942；
color：＃ffffff；
}
＃students tr. alt td {
color：＃000000；
```

```
background-color：#EAF2D3；
}
```

2. resources/js 目录。

left.js

```
$(document)
.ready(
function(){
$('.Left_Bar_title').click(
function(){
$(this).parent().find('p').slideToggle();
$(this).parent().siblings('li').find('p').slideUp();
});
});
```

参考文献

[1]埃克尔，埃克尔，陈吴鹏. Java 编程思想[M]. 机械工业出版社，2007.

[2]高洪岩. Java EE 核心框架实战[M]. 人民邮电出版社，2014.

[3]耿祥义，张跃平. JSP 程序设计[M]. 清华大学出版社，2015.

[4]贾蓓，镇明敏，杜磊. Java Web 整合开发实战：基于 Struts 2＋Hibernate＋Spring[M]. 清华大学出版社，2013.

[5]克朗斯帕斯卡雷洛詹姆斯. Ajax 实战[M]. 人民邮电出版社，2006.

[6]BudiKurniawan，克尼亚万，崔毅，等. Servlet 和 JSP 学习指南[M]. 机械工业出版社，2013.

[7]李刚. 轻量级 Java EE 企业应用实战：Struts 2＋Spring 3＋Hibernate 整合开发[J]. 2014.

[8]QST 青软实训. Java EE 轻量级框架应用与开发[M]. 清华大学出版社，2016.

[9]GeorgeReese，Reese，石永鑫，等. JDBC 与 Java 数据库编程[M]. 中国电力出版社，2002.

[10]佟强. JSP 程序设计[M]. 清华大学出版社，2013.